T0192770

COSMOLOGY FOR PHYSICISTS

Series in Astronomy and Astrophysics

The *Series in Astronomy and Astrophysics* includes books on all aspects of theoretical and experimental astronomy and astrophysics. Books in the series range in level from textbooks and handbooks to more advanced expositions of current research.

Series Editors:
M Birkinshaw, University of Bristol, UK
J Silk, University of Oxford, UK
G Fuller, University of Manchester, UK

Recent books in the series

Cosmology for Physicists
David Lyth

Stellar Explosions: Hydrodynamics and Nucleosynthesis
Jordi José

Fundamentals of Radio Astronomy: Observational Methods
Jonathan M Marr, Ronald L Snell and Stanley E Kurtz

Astrobiology: An Introduction
Alan Longstaff

An Introduction to the Physics of Interstellar Dust
Endrik Krugel

Numerical Methods in Astrophysics: An Introduction
P Bodenheimer, G P Laughlin, M Rózyczka, H W Yorke

Very High Energy Gamma-Ray Astronomy
T C Weekes

The Physics of Interstellar Dust
E Krügel

Dust in the Galactic Environment, 2nd Edition
D C B Whittet

Dark Sky, Dark Matter
J M Overduin and P S Wesson

An Introduction to the Science of Cosmology
D J Raine and E G Thomas

The Origin and Evolution of the Solar System
M M Woolfson

The Physics of the Interstellar Medium, 2nd Edition
J E Dyson and D A Williams

Series in Astronomy and Astrophysics

COSMOLOGY FOR PHYSICISTS

David Lyth

Lancaster University, United Kingdom

CRC Press
Taylor & Francis Group
Boca Raton London New York

CRC Press is an imprint of the
Taylor & Francis Group, an **informa** business

CRC Press
Taylor & Francis Group
6000 Broken Sound Parkway NW, Suite 300
Boca Raton, FL 33487-2742

First issued in paperback 2020

© 2017 by Taylor & Francis Group, LLC
CRC Press is an imprint of Taylor & Francis Group, an Informa business

No claim to original U.S. Government works

ISBN-13: 978-0-367-57438-3 (pbk)
ISBN-13: 978-1-4987-5531-3 (hbk)

Visit the Taylor & Francis Web site at
http://www.taylorandfrancis.com

and the CRC Press Web site at
http://www.crcpress.com

To my dear wife Margaret

Contents

SECTION II The Big Bang

Preface

Cosmology is the study of the large-scale properties of the Universe and its main focus is the Universe's history. As an observational science, the subject was born with discovery by Edwin Hubble in the 1920s, that our galaxy is one of many and that the galaxies are moving apart. Since each galaxy has a fixed size, that suggests the existence at early times of a Universe without galaxies and by the same token it suggests the existence of a Universe without any macroscopic objects at all. A Universe consisting simply of a more or less uniform expanding gas.

Such a scenario was confirmed by the discovery in 1964 of the cosmic microwave background (CMB) The CMB is almost exactly isotropic, reflecting the near homogeneity of the Universe when it was emitted. Even more interesting is the slight anisotropy of the CMB, that was observed in 1992. The CMB anisotropy, supplemented by a host of other observations, has allowed cosmologists to determine beyond reasonable doubt the history of the Universe after the first hundredth of a second. It has also led to wide acceptance, as the only game in town, of the hypothesis that the initial condition for the Universe's evolution was set during an early era of cosmic inflation (so called because the rate of expansion was then increasing).

This book gives an account of cosmology for those who know physics at the level of a late undergraduate student. It should be useful to practically anybody who is, or has been, such a student. The only exceptions would be those actively engaged in early-Universe research, numbering I suppose a few hundred at the present time.

With these things in mind, the book is divided into three parts. Part I deals with some more physics. There is a review of particle physics including a discussion of neutrino masses, and there is an account of General Relativity. Part II is an account of the known history, and Part III deals with inflation.

My understanding of cosmology has largely been acquired during research projects. Among many collaborators I should like to mention in particular Andrew Liddle, Antonio Riotto, Yeinzon Rodriguez and David Wands.

I

Physics

Particle physics

CONTENTS

At the present time, our understanding of physics is in two parts. There is Einstein's General Relativity which describes gravity, and the rest of physics which ignores gravity and uses the description of spacetime provided by Special Relativity. At the fundamental level the rest of physics consists of a description of particles and their interactions, which is largely though not entirely provided by what is called the Standard Model. After some preliminaries, this chapter describes the current status of particle physics.

1.1 SPECIAL RELATIVITY

In this section I recall some results of Special Relativity, setting $c = 1$. Special Relativity invokes preferred spacetime coordinates, called Minkowski coordinates, in which the laws of physics become particularly simple. These coordinates define an inertial frame. In terms of Minkowski coordinates, the spacetime interval is given by

$$ds^2 = -dt^2 + dx^2 + dy^2 + dz^2. \tag{1.1}$$

(Some authors multiply the right hand side by -1.) Denoting the coordinates by $\{x^0, x^1, x^2, x^3\}$ it is convenient to write

$$ds^2 = \eta_{\mu\nu} dx^\mu dx^\nu, \tag{1.2}$$

where $\eta_{\mu\nu}$ is diagonal with elements $(-1, 1, 1, 1)$, and identical pairs of upper and lower indices are summed over even though there is no summation sign. This is the summation convention which I will use from now on.

A Lorentz transformation takes us from one set of Minkowski coordinates to another, with the origin fixed. The most general Lorentz transformation can be built from rotations, and a Lorentz boost along the x-axis. The Lorentz boost is the change seen by an observer moving with speed v along the x-axis and it given by

$$x' = \gamma(x - vt), \qquad t' = \gamma(t - vx) \qquad (1.3)$$
$$y' = y, \qquad z' = z, \qquad (1.4)$$

where $\gamma = 1/\sqrt{1 - v^2}$.

Reverting to coordinates x^μ, a 4-vector A^μ is a quantity which transforms like x^μ under Lorentz transformation. The quantity

$$\eta_{\mu\nu} A^\mu A^\nu \equiv A^\mu A_\mu \qquad (1.5)$$

is Lorentz-invariant where in the second expression $A_\mu \equiv \eta_{\mu\nu} A^\nu$. I will call A^μ an upper-index 4-vector and A_μ a lower-index 4-vector. (They are also called contravariant and covariant 4-vectors.) As in this case, the four spacetime indices will always be denoted collectively by a Greek letter. The three spatial indices will be denoted collectively by a Roman letter. For them, there is no difference between the upper and lower placement because $\eta_{ij} = 1$.

The motion of an object corresponds to a path through spacetime, called the worldline of the object. The energy-momentum 4-vector p^μ of the particle has components (E, p^1, p^2, p^3) where E is the energy and p^i are the components of the momentum. The mass m of the object is defined by

$$-p^\mu p_\mu \equiv E^2 - p^2 = m^2. \qquad (1.6)$$

If m is nonzero we can go to the rest frame, where the momentum is zero and $E = m$. If $m = 0$, as for a photon, we have instead $E = p$.

1.2 STANDARD MODEL

The Standard Model was formulated in the late 1970's after a series of dramatic break-throughs. As the name implies, it was regarded as a working hypothesis and its continuing success was hardly anticipated.

The elementary particles of the Standard Model are listed in Table A.5. They are the six quarks u, d, c, s, t and b and the six leptons e (electron), μ, τ, ν_e, ν_μ and ν_τ. These come with anti-particles which are usually denoted by $\bar{u}$, $\bar{d}$ and so on (only the $\bar{e}$ has a special name, the positron). There are also four gauge bosons; the photon denoted by γ, the gluon, the Z and the W$^+$ which have spin 1. The photon and Z have no anti-particle. The W$^+$ has the anti-particle W$^-$ and there is also an anti-gluon. Finally, there is the Higgs boson with spin 0 and no anti-particle.

Each quark comes in three versions, and the gluon comes in eight versions, the versions being distinguished by the quantum number called colour. That can be ignored for most purposes though, because the observable quarks and the observable gluon correspond to quantum state vectors which are superpositions of state vectors whose total colour is zero.

The composite particles of the Standard Model are called hadrons. With rare exceptions, the hadrons are either baryons or mesons. According to the usual description, a baryon is made out of three quarks. In particular, the proton consists of uud and the neutron consists of udd. A meson is made out of a quark and an anti-quark; for instance a π^+ consists of $u\bar{d}$. Each baryon comes with an anti-baryon, and each meson comes with an anti-meson unless its quark and anti-quark are of the same species. The exceptional hadrons are of two kinds. First, there can be more than three quarks; at least one tetraquark has been observed consisting of a two quarks and two anti-quarks, and at least one pentaquark consisting of four quarks and an anti-quark.

These, the usual descriptions of the composite particles, are actually over-simplified because in addition to the quarks mentioned each composite particle contains an indefinite number of quark-anti-quark pairs and an indefinite number of gluons. ('Indefinite number' here means that the quantum state is not an eigen-state of the number operator.) The Standard Model actually predicts the existence of what are called glueballs, with no quark content at all, but no glueball has been identified with certainty.

Each particle has a spin, defined as the maximum value in units of $\hbar$ of the component of angular momentum in a given direction. The direction is conveniently chosen to be the direction of motion, and the component of angular momentum in that direction is called the helicity. As with any angular momentum component, quantum physics requires that the spin is an integer or half-integer. The quarks and leptons have spin 1/2, as do the proton and neutron. The gauge bosons have spin 1 and the Higgs has spin 0.

For a massive particle, we can identify the possible helicities by going to the rest frame. For the quarks and charged leptons this gives helicities $\pm 1/2$. (The case of neutral leptons, ie. neutrinos will be discussed in Section 1.6.) For the Z and $W_\pm$ it gives helicities 1, 0 or -1. For the photon and gluon there is no rest frame and quantum physics gives just two helicities ± 1. For the photon these correspond to right- and left circularly polarised electromagnetic waves.

The Standard Model is a quantum field theory, which means that each particle species corresponds to the oscillation of some operator called a field.[1] Most of the fields though, don't correspond to observables because they are not self-adjoint. The exceptions are the Higgs field which corresponds to the Higgs boson, and the gauge fields which correspond to the gauge bosons.[2] The gauge field corresponding to the photon is the 4-vector potential A^μ that specifies the electromagnetic field.

Although the Standard Model accounts for almost all observations it's not perfect. It doesn't account for the neutrino masses. It doesn't include any particle species that could be the Cold Dark Matter (CDM), which as we shall see is the dominant component of the Universe. Nor does it include a field that could

---

[1] That is in the Heisenberg picture, where the quantum state is time independent and the operators are time dependent.

[2] The gauge fields corresponding to the $W_\pm$ are actually complex, and for them it is the real and imaginary parts that correspond to Hermitian operators.

be responsible for inflation, at least if General Relativity applies then.[3] For these and other reasons, various extensions of the Standard Model have been proposed. The extensions use the same general framework as the Standard Model, namely quantum field theory. They usually invoke additional particle species, which have specified interactions with each other and with the Standard Model particles.

1.3 COLLISION AND DECAY PROCESSES

A particle can disappear, to be replaced by two or more different particles. That is a decay process. Also, two particles may collide. That is a collision process which may have a number of outcomes. The particles may simply bounce off each other (elastic scattering), or else new particles may be created. In the latter case, the incoming particles may or may not survive.

Instead of involving particles, a collision or decay process can involve ions, atoms or molecules. Collision and decay processes of all kinds are important in cosmology. They are important in themselves, simply because they occur, and they are also important because they allow astronomers to make observations.

Collision and decay processes are also important here on Earth, in many contexts. From a fundamental viewpoint, what are and have been most important are collision processes, engineered in the laboratory with the object of creating new particles and investigating their interactions.[4] The machines used for that purpose are of two kinds. One kind accelerates a beam of particles so that it has high energy, and bounces the beam off a fixed target. That is simply called an accelerator. Another kind accelerates two beams and collides them head-on. That is called a collider.

The ability of an accelerator or collider to create new particles is determined by the energy of the collision in the centre-of-mass frame. This is because the energy in that frame has to be at least equal to the rest energy of the created particle. The advance in our knowledge has come from the building of accelerators and colliders with ever-increasing energy. The leading machine at present is the LHC at CERN (Geneva) which is colliding protons at the time of writing (April 2016) with centre-of-mass energy 6.5 TeV. The LHC created the Higgs boson for the first time. It has confirmed the Standard Model, except that a new particle has been found with rest energy 750 GeV. The status of the new particle is not yet clear.

1.4 CONSERVED QUANTITIES

Every collision and decay process conserves energy and momentum, and according to the Standard Model it also conserves two other quantities, namely electric charge Q and baryon number B. In addition, the collision processes that occur during the known history of the Universe, and at colliders, conserve to high accuracy the three

---

[3]If it does not, the gravity theory that replaces it might allow the Higgs field to be responsible for inflation.

[4]Instead of particles one may deal with ions.

lepton numbers L_e. L_μ and L_τ.[5] Table A.6 of Appendix A.3 shows the charge, baryon number and lepton numbers carried by the elementary particles, and by the proton and neutron (nucleons).

Baryon number and the lepton numbers are just book-keeping devices. Their conservation tell us that some processes don't happen, and that's all. For instance, a proton can't decay according to the Standard Model because it is the lightest particle with non-zero baryon number. In contrast, electric charge has significance going beyond mere conservation. The electric charge of an object determines the strength of the electromagnetic field that it generates, and the effect on its motion of that field.

Energy conservation places important restrictions on collision and decay process. To understand them, we have to remember that the energy of a particle with mass m and speed v is

$$E = \frac{mc^2}{\sqrt{1 - (v/c)^2}}. \tag{1.7}$$

For a particle at rest $E = mc^2$ (the rest energy) while for a particle in motion E is bigger, the difference being the kinetic energy.

Working in the rest frame of a decaying particle, the initial energy is equal to the rest energy of the decaying particle. The final energy is equal to the sum of the rest energies of the final particles plus their total kinetic energy. It follows that a decay can occur, only if sum of the rest energies of the decay products is less than the rest energy of the decaying particle. In other words, if the total mass of the decay products is less than the mass of the decaying particle.

Collision processes, from the viewpoint of energy conservation, are of two kinds depending on whether the total rest mass of the incoming particles is smaller or bigger than the total rest mass of the produced particles. If it is smaller, the incoming particles must have enough kinetic energy to make up the difference. If instead it is bigger, the process can occur no matter how small is the kinetic energy of the incoming particles. In the latter case, the collision process creates kinetic energy. Exactly the same thing is true if instead of particles we have nuclei, atoms or molecules. If we deal with atoms or molecules we are talking about a chemical reaction. Then the former case, where energy must be supplied, is called an endothermic reaction while the latter case, which creates energy is called an exothermic reaction. An endothermic reaction cools down the substance in which it is taking place, while an exothermic reaction heats it up.

---

[5]Neutrino oscillations violate the conservation of the separate lepton numbers. Also, collisions with sufficiently high energy will, according to the Standard Model, conserve only the differences between the lepton numbers together with $B - L$ where L is the sum of the lepton numbers.

1.5 NATURAL UNITS

In Section 1.1 I set $c = 1$. It is often convenient to go further and set both c and $\hbar$ equal to 1. Taking the remaining mechanical unit to be energy we arrive at what are called natural units.

With natural units, dimensional analysis becomes much simpler because there is just one dimension, namely energy E. Let us see how to determine the dimension of a quantity in natural units, of a quantity whose dimension in ordinary units is $M^p L^q T^r$. First, since c is a speed it has ordinary dimensions LT^{-1} and since we are setting it to 1 we can replace L by T so that the dimension of our quantity becomes $M^p T^{q+r}$. Also, since the rest energy of an object with mass m is mc^2 we can replace M by E so that the dimension becomes $E^p T^{q+r}$. Finally, since a system with energy E has a wave function which is proportional to $\exp(i\omega t)$ with $E = \hbar\omega$, it follows that $\hbar$ has the dimension ET.[6] Since we are setting $\hbar = 1$, we can replace T by E^{-1} to obtain the final dimension E^{p-q-r}.

As an example of the use of this formula, let us consider energy density. In ordinary units energy has dimension ML^2T^{-2} (remember $E = mc^2$) which means that energy density has dimension $M^2 L^{-1} T^{-2}$. This corresponds to $p = 1$, $q = -1$ and $r = -2$ which means that the dimension of energy density in natural units is E^{1+1+2} or E^4.

1.6 NEUTRINO MASSES

Neutrino masses may be important for cosmology. In this section I discuss them using natural units.

Our knowledge of neutrino masses comes from what are called neutrino oscillations. Neutrino oscillations are observed for cosmic ray neutrinos, neutrinos produced in colliders, and neutrinos emerging from the sun. The following account applies to the first two cases where the neutrino can be taken to be travelling through the vacuum, but it is only slightly modified for neutrinos travelling through the sun.

The states $|\alpha\rangle$ ($\alpha =$ e, μ or τ) of a neutrino produced in a collision decay process are superpositions of the states $|i\rangle$ ($i = 1$, 2 or 3) that have definite mass. We can therefore define a unitary mixing matrix U by writing

$$|i\rangle = \sum_{\alpha} U_{i\alpha} |\alpha\rangle. \tag{1.8}$$

In the Schrödinger picture

$$|i, \tau\rangle = e^{-im_i\tau} |i, 0\rangle, \tag{1.9}$$

where τ is the time in the neutrino's rest frame. In terms of lab frame quantities,

$$m_i \tau_i = E_i t - p_i L, \tag{1.10}$$

---

[6]There are of course other equations that could be used to determine the dimension of $\hbar$.

where E_i and p_i are the energy and momentum of a neutrino with definite mass and L is the distance it travels in time t.[7] Observed neutrinos have speed close to c and hence $p_i \gg m_i$, which gives

$$m_i \tau_i \simeq (E_i - p_i)L \simeq \frac{m_i^2 L}{2E_i}, \qquad (1.11)$$

where the last equality follows from $E_i = \sqrt{p_i^2 + m_i^2}$. Using Eq. (1.8) and its inverse, the state of a neutrino when it has travelled a distance L is

$$|\alpha, L\rangle = \sum_{\beta, i} \left[U_{\alpha i}^* U_{\beta i} e^{-i(m_i^2 L / 2E_i)} \right] |\beta, 0\rangle. \qquad (1.12)$$

The modulus-squared of the square bracket is the transition probability, which is observable. This gives some information about the mixing matrix, and it determines the mass differences. Present observation gives the information

$$|m_3^2 - \tfrac{1}{2}(m_1^2 + m_2^2)| = (2.40 \pm 0.08) \times 10^{-3} \, \text{eV}^2 \qquad (1.13)$$

$$m_2^2 - m_1^2 = (7.5 \pm 0.3) \times 10^{-5} \, \text{eV}^2. \qquad (1.14)$$

If the masses are well separated this gives $m_3 = 0.05\,\text{eV}$, $m_2 = 0.009\,\text{eV}$ and $m_1 \ll m_2$. If two masses are almost equal it gives only $m_1 \simeq m_2 \simeq 0.05\,\text{eV}$ and $m_3 \ll 0.05\,\text{eV}$. If all three are almost equal neutrino oscillation gives no information about the sum of the masses, but an upper bound on that sum is implied by the Cosmic Microwave Background (CMB) anisotropy described in Chapter 14 and the baryon acoustic oscillation described in Chapter 13. Taken together these give $\sum_j m_j < 0.23\,\text{eV}$. If the bound is well-satisfied as seems likely from the preceding analysis, neutrino masses have no significant effect on cosmology.

1.7 NEUTRINOS AT REST

The energy of neutrinos produced in collisions is always much bigger than their rest energy, which means that they travel with speed close to c. Within the accuracy of observation, the component of spin along the direction of motion (helicity) is always $-\hbar/2$ for neutrinos and $+\hbar/2$ for anti-neutrinos.

Current observations give no information about the nature of the neutrino and anti-neutrino rest. There are two possibilities for that situation. One is that the neutrino and anti-neutrino really are the anti-particles of each other, as the names imply. Spin-half particles that come with an anti-particle are called Dirac particles. In that case, there would be two possible helicities for a neutrino with speed close to c, and two for an anti-neutrino. Such a thing is allowed by the Standard Model, because it predicts that only one of the helicities would be created with significant probability in collision and decay processes. The other possibility is that what we

---

[7] The right hand side is $p_\mu x^\mu$, where p^μ is the 4-momentum and x^μ is the spacetime position.

call the anti-neutrino turns out to be actually a neutrino when viewed in its rest frame, so that there is in reality no anti-neutrino. Spin-half particles that have no anti-particle are called Majorana particles. The first of these two possibilities would be ruled out by the observation of the neutrino-less double beta decay process shown in Figure 1.1.

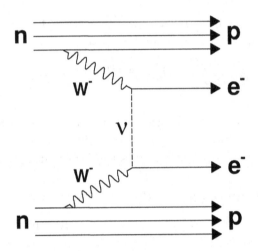

Figure 1.1 The Feynman diagram for neutrino-less double beta decay. which is allowed if the neutrino is a Majorana particle. I thank Alex Finch for supplying this diagram.

EXERCISES

1. In natural units, what are the dimensions of length and time?

2. By inserting the appropriate powers of c and $\hbar$, verify the conversion factors given in Table A.1.

3. How many species of elementary particle are there, according to the Standard Model?

4. Which conservation law forbids the decay of the electron?

5. Using Table A.5, calculate the mass of a nucleon (proton or neutron) in kilograms. Estimate the number density of nucleons in the Earth, with mass 6.0×10^{24} kg and radius 6400 m. Using Table A.3 and Eq. (8.17), calculate the collision rate Γ for a neutrino with energy E, and the value of E in eV for which Γ^{-1} would be equal to the time taken for the neutrino to pass through the Earth. (Neutrinos created in collision and decay processes actually have far smaller energies which means that a typical neutrino passes through the Earth without interacting.)

Curved spacetime

CONTENTS

Special Relativity ignores gravity. In the presence of gravity we have to replace Special Relativity by General Relativity, and then it is impossible to find Minkowski coordinates, such that the spacetime interval is given by Eq. (1.1). One says that spacetime in that case is curved, while in the absence of gravity spacetime is flat.[1] In this chapter I describe curved spacetime.

2.1 SPACETIME METRIC

According to General Relativity, there is an object $g_{\mu\nu}$, called the spacetime metric, in terms of which the interval is

$$ds^2 = g_{\mu\nu}dx^\mu dx^\nu. \tag{2.1}$$

where $g_{\mu\nu}$ is called the metric of spacetime. The metric depends on the spacetime position. Eq. (2.1) is called the line element, and by specifying it one specifies the metric. (Eq. (2.1) of course makes sense, only after ds^2 has been defined. I will show how that is done shortly.)

We assume that at any *one* point in spacetime, we can find coordinates such that

$$g_{\mu\nu} = \eta_{\mu\nu}. \tag{2.2}$$

I will call these locally orthonormal coordinates. Going from locally orthonormal coordinates to some generic coordinates x'^α, we will have

$$dx^\mu = \frac{\partial x^\mu}{\partial x'^\alpha}dx'^\alpha. \tag{2.3}$$

---

[1]This is because of the analogy with surfaces. On a flat surface one can find Cartesian coordinates, but on a curved surface that is generally impossible.

At the chosen point in spacetime we will therefore have

$$g'_{\alpha\beta} = \frac{\partial x^\mu}{\partial x'^\alpha}\frac{\partial x^\nu}{\partial x'^\beta}\eta_{\mu\nu} \tag{2.4}$$

where the prime on $g'_{\alpha\beta}$ reminds us that it is evaluated using the primed coordinates. This relation restricts the form of $g_{\mu\nu}$, and in particular requires $g_{\mu\nu} = g_{\nu\mu}$.[2]

We can also define $g^{\mu\nu}$ as the object that raises indices. It is the inverse of $g_{\mu\nu}$ because for any vector A^μ we have

$$A^\mu = g^{\mu\nu}A_\nu = g^{\mu\alpha}\left(g_{\alpha\beta}A^\beta\right), \tag{2.5}$$

and since this is true for any A^μ it requires $g^{\mu\alpha}g_{\alpha\nu} = \delta^\mu_\nu$. Going from any coordinates x^μ to new coordinates x'^α, we learn from Eq. (2.3) that

$$g'_{\alpha\beta} = \frac{\partial x^\mu}{\partial x'^\alpha}\frac{\partial x^\nu}{\partial x'^\beta}g_{\mu\nu}. \tag{2.6}$$

2.2 FOUR-VECTORS AND TENSORS

At a given point in spacetime, a 4-vector A^μ is defined as an object that transforms in the same way as dx^μ;

$$A'^\mu = \frac{\partial x'^\mu}{\partial x^\alpha}A^\alpha. \tag{2.7}$$

This is the same as the definition of a 4-vector in Special Relativity, except that there the transformation is taken to go from one set of Minkowski coordinates to another. That makes $\partial x'^\mu/\partial x^\alpha$ independent of position and it means that the definition of a 4-vector does not refer to any particular point in spacetime. In General Relativity, the definition of a 4-vector refers to a particular point in spacetime.

Instead of A^μ we can consider $A_\mu = g_{\mu\nu}A^\nu$. Using Eq. (2.6) one finds that its transformation is

$$A'_\mu = \frac{\partial x^\alpha}{\partial x'^\mu}A_\alpha. \tag{2.8}$$

We also have $A^\mu = g^{\mu\nu}A_\nu$ where $g^{\mu\nu}$ is the matrix inverse of $g_{\mu\nu}$.

A second-rank tensor $B^{\mu\nu}$ is defined as a quantity which transforms like a product of two upper-index vectors;

$$B'^{\mu\nu} = \frac{\partial x^\mu}{\partial x'^\alpha}\frac{\partial x^\nu}{\partial x'^\beta}B'^{\alpha\beta}. \tag{2.9}$$

One or both of the indices can be lowered by the action of $g_{\mu\nu}$, giving objects with the transformations

$$B'^\mu_{\ \nu} = \frac{\partial x'^\mu}{\partial x^\alpha}\frac{\partial x^\beta}{\partial x'^\nu}B^\alpha_{\ \beta} \tag{2.10}$$

$$B'_{\mu\nu} = \frac{\partial x'^\alpha}{\partial x^\mu}\frac{\partial x'^\beta}{\partial x^\nu}B_{\alpha\beta}. \tag{2.11}$$

---

[2]As in this case, the indices at a given point in the text are chosen without regard to the choice at previous points. For Greek letters one generally chooses μ and ν if only two are involved, and for Roman letters i and j.

We see that in each case, the transformation is the one suggested by the placement of the indices. We see from Eq. (2.6) that the metric tensor is indeed a tensor according to this definition. A lower index can be raised again by the action of the inverse matrix $g^{\mu\nu}$.

All of this can be repeated for a tensor of higher rank, involving more than two indices. Going the other way, we can consider a scalar, which carries no indices and is the same in every coordinate system.

2.3 COVARIANT DERIVATIVE AND D'ALEMBERTIAN

Now I consider a quantity that is defined within some region of spacetime and is smoothly varying. To make equations easier to read, I will use the compact notation $\partial_\mu \equiv \partial/\partial x^\mu$.

If the quantity is a scalar, denoted by ϕ, $\partial_\mu \phi$ is a 4-vector because it has the correct transformation (2.8). But if A^μ is a 4-vector defined throughout spacetime, the ordinary derivative $\partial_\nu A^\mu$ is not a tensor because it has the wrong transformation. The covariant derivative $D_\nu A^\mu$ is defined as the second-rank tensor which reduces to the ordinary derivative in a locally inertial frame at each point. If A'^α and x'^α refer to a locally inertial frame, while A^μ and x^μ refer to generic coordinates, the tensor transformation (2.10) and the chain rule give

$$D_\nu A^\mu = \frac{\partial x^\mu}{\partial x'^\beta}\frac{\partial x'^\alpha}{\partial x^\nu}\frac{\partial A'^\beta}{\partial x'^\alpha} = \frac{\partial x^\mu}{\partial x'^\beta}\partial_\nu A'^\beta. \tag{2.12}$$

Using Eq. (2.7) for A'^β this can be written

$$D_\nu A^\mu = \partial_\nu A^\mu + \Gamma^\mu_{\nu\alpha} A^\alpha, \tag{2.13}$$

where $\Gamma^\mu_{\alpha\beta}$, called the Levi-Civita connection, is given by

$$\Gamma^\mu_{\alpha\beta} = \frac{\partial^2 x'^\lambda}{\partial x^\alpha \partial x^\beta}\frac{\partial x^\mu}{\partial x'^\lambda}. \tag{2.14}$$

Repeating all this for a lower-index vector A_μ one finds

$$D_\nu A_\mu = \partial_\nu A^\mu - \Gamma^\alpha_{\nu\mu} A_\alpha. \tag{2.15}$$

The effect of D_μ on a tensor can be calculated by remembering that it transforms as a product of 4-vectors. For a product $A_\mu B_\mu$ we would have[3]

$$
\begin{aligned}
D_\mu (A_\alpha B_\beta) &= (D_\mu A_\alpha) B_\beta + A_\alpha (D_\mu B_\beta) & (2.16)\\
&= \partial_\mu (A_\alpha B_\beta) - \Gamma^\nu_{\mu\alpha} A_\nu B_\beta - \Gamma^\nu_{\mu\beta} A_\alpha B_\nu. & (2.17)
\end{aligned}
$$

---

[3]The first line is obviously true in a locally inertial frame, and it is preserved by a coordinate transformation.

The covariant derivative of the metric tensor vanishes because $g_{\mu\nu}$ in a locally inertial frame has the constant value $\eta_{\mu\nu}$. It follows that

$$D_\mu(g_{\nu\lambda}) = \frac{\partial g_{\nu\lambda}}{\partial x^\mu} - \Gamma^\alpha_{\mu\nu} g_{\alpha\lambda} - \Gamma^\alpha_{\mu\lambda} g_{\nu\alpha} = 0. \tag{2.18}$$

After some manipulation, this gives an expression for the Levi-Cevita connection in terms of the metric tensor;[4]

$$\Gamma^\mu_{\alpha\beta} = \frac{1}{2} g^{\mu\nu} \left(\frac{\partial g_{\nu\alpha}}{\partial x^\beta} + \frac{\partial g_{\nu\beta}}{\partial x^\alpha} - \frac{\partial g_{\alpha\beta}}{\partial x^\nu} \right). \tag{2.19}$$

The Levi-Civita connection is not a tensor because it does not have the appropriate transformation. Nevertheless, it is useful to define

$$\Gamma_{\mu\nu\lambda} = g_{\lambda\kappa} \Gamma^\kappa_{\mu\nu} \tag{2.20}$$

$$= \left(\frac{\partial g_{\nu\alpha}}{\partial x^\beta} + \frac{\partial g_{\nu\beta}}{\partial x^\alpha} - \frac{\partial g_{\alpha\beta}}{\partial x^\nu} \right), \tag{2.21}$$

which satisfies

$$\partial_\lambda g_{\mu\nu} = \Gamma_{\mu\nu\lambda} + \Gamma_{\nu\mu\lambda}. \tag{2.22}$$

Along a parameterised line $x^\mu(\lambda)$, the rate of change of a vector A^μ is given by

$$\frac{DA^\mu}{D\lambda} = \frac{dx^\nu}{d\lambda} D_\nu A^\mu \tag{2.23}$$

$$= \frac{dx^\nu}{d\lambda} \left[\frac{\partial A^\mu}{\partial x^\nu} + \Gamma^\mu_{\nu\alpha} A^\alpha \right] \tag{2.24}$$

$$= \frac{dA^\mu}{d\lambda} + \Gamma^\mu_{\nu\alpha} A^\alpha \frac{dx^\nu}{d\lambda}. \tag{2.25}$$

In flat spacetime the D'Alembertian $\Box$ acting on a scalar ϕ is defined by $\Box\phi \equiv \eta^{\mu\nu} \partial_\mu \partial_\nu \phi$. In curved spacetime it is defined by

$$\Box\phi \equiv g^{\mu\nu} D_\mu \partial_\nu \phi. \tag{2.26}$$

It is given by the convenient expression

$$\Box\phi = \frac{1}{\sqrt{-g}} \partial_\mu \left(\sqrt{-g} g^{\mu\nu} \partial_\nu \phi \right), \tag{2.27}$$

where g is the determinant of $g_{\mu\nu}$. To arrive at this expression one needs the formulas

$$g = \sum_\nu g_{\mu\nu} \Delta^{\mu\nu} \qquad \text{any fixed } \mu \tag{2.28}$$

$$g^{\mu\nu} = \Delta^{\mu\nu}/g \tag{2.29}$$

---

[4]To arrive at this result, cyclically permute the indices to obtain two more expressions. Now, remembering that $\Gamma^\mu_{\nu\alpha} = \Gamma^\mu_{\alpha\nu}$, add two of the expressions and subtract the third, to obtain $2g_{\alpha\nu}\Gamma^\nu_{\beta\mu} = \partial_\mu g_{\alpha\beta} + \partial_\beta g_{\alpha\mu} - \partial_\alpha g_{\beta\mu}$ where $\partial_\mu \equiv \partial/\partial x^\mu$. Contracting this expression with $g^{\gamma\alpha}$ gives the desired result.

where $\Delta^{\mu\nu}$ is the cofactor of the element $g_{\mu\nu}$.[5] Up to a sign, the cofactor of an element of the matrix $g_{\mu\nu}$ is defined as the determinant of the matrix that is obtained by deleting the row and column containing that element. The cofactor $\Delta^{\mu\nu}$ therefore contains no factor $g_{\mu\nu}$ and we have using Eqs. (2.28), (2.29), and (2.22)

$$\partial_\lambda g = \frac{\partial g}{\partial g_{\mu\nu}}\partial_\lambda g_{\mu\nu} \tag{2.30}$$

$$= \Delta^{\mu\nu}\partial_\lambda g_{\mu\nu} \tag{2.31}$$

$$= g g^{\mu\nu}\partial_\lambda g_{\mu\nu} \tag{2.32}$$

$$= g g^{\mu\nu}\left(\Gamma_{\mu\nu\lambda}+\Gamma_{\nu\mu\lambda}\right) \tag{2.33}$$

$$= g\left(\Gamma^\mu_{\mu\lambda}+\Gamma^\nu_{\nu\lambda}\right) \tag{2.34}$$

$$= 2g\Gamma^\mu_{\mu\lambda}. \tag{2.35}$$

This gives

$$\Gamma^\mu_{\mu\nu} = \frac{1}{2}g^{-1}\partial_\nu g = \frac{1}{\sqrt{|g|}}\partial_\nu g. \tag{2.36}$$

Using Eqs. (2.15) and (2.26) gives Eq. (2.27).

2.4 EQUIVALENCE PRINCIPLE

At a point where Eq. (2.2) applies, one can find coordinates such that[6]

$$\partial g_{\mu\nu}/\partial x^\tau = 0. \tag{2.37}$$

This is called the local flatness theorem. To prove it, we need to show that it is possible to go from coordinates in which Eq. (2.2) applies but Eq. (2.37) may not, to new coordinates where they both apply. Denoting the original coordinates by x^μ and the new ones by x'^α, we have at the point where Eq. (2.2) applies

$$0 = \frac{\partial g'_{\mu\nu}}{\partial x'^\tau} \tag{2.38}$$

$$= \frac{\partial}{\partial x'^\tau}\frac{\partial x^\alpha}{\partial x'^\mu}\frac{\partial x^\beta}{\partial x'^\nu}g_{\alpha\beta} \tag{2.39}$$

$$= \frac{\partial^2 x^\alpha}{\partial x'^\tau \partial x'^\mu}\frac{\partial x^\beta}{\partial x'^\nu}g_{\alpha\beta} + \frac{\partial x^\alpha}{\partial x'^\mu}\frac{\partial^2 x^\beta}{\partial x'^\tau \partial x'^\nu}g_{\alpha\beta} + \frac{\partial x^\alpha}{\partial x'^\mu}\frac{\partial x^\beta}{\partial x'^\nu}\frac{\partial g_{\alpha\beta}}{\partial x'^\tau}. \tag{2.40}$$

These are $4\times 10 = 40$ relations that can indeed be satisfied by choosing the $4\times 10 = 40$ quantities $\partial^2 x^\alpha/\partial x'^\tau \partial x'^\mu$. (The numbers are 4×10 and not 4×16, because in both cases there is symmetry under the interchange of a pair of indices.)

Coordinates in which both Eq. (2.2) and Eq. (2.37) apply are called locally

---

[5]$\Delta^{\mu\nu}$ is not a tensor because g is not a scalar, i.e. is not invariant under a change of coordinates. I use upper indices for $\Delta^{\mu\nu}$ only to make the equations look neat.

[6]This can be done whether or not Eq. (2.2) is valid but we are interested in the case that it is valid.

inertial coordinates, defining what is called a locally inertial frame. Fundamental laws of physics that do not include gravity can be formulated in the context of Special Relativity. According to what is called the equivalence principle, these laws apply in every locally inertial frame.[7] In other words, gravity is abolished in a locally inertial frame. This then is how the spacetime interval ds^2 appearing in Eq. (2.1) is defined; it is defined by Eq. (1.1) using the Minkowski coordinates of Special Relativity in a locally inertial frame.

After invoking the equivalence principle we can go, at each point in spacetime, from a locally inertial frame to a generic coordinate system. We then have a formulation of the laws of physics which applies even in the presence of gravity.

For the equivalence principle to work, it's important that the fundamental laws of physics in Special Relativity involve only second spacetime derivatives of scalars and only first spacetime derivatives of 4-vectors and tensors.[8] This allows them (using the Levi-Civita connection) to be written in terms of $g_{\mu\nu}$, and its *first* spacetime derivatives which vanish in a locally inertial frame. If higher spacetime derivatives were involved, the laws would cease to have a unique form in a locally inertial frame because they would involve higher spacetime derivatives of the metric tensor.

Since the laws of physics in General Relativity can be written in terms of the metric, they take the same form in every coordinate system. This is the reason for the term "General". (The laws of physics in Special Relativity also take the same form in every coordinate system if we invoke the covariant derivative, but there we can instead go to Minkowski coordinates and just use the ordinary derivative. In General Relativity the covariant derivative is unavoidable and there is no preferred coordinate system.)

In this discussion I have used the strict definition of term "locally inertial frame", which refers to a single spacetime point. In practice one takes the term to apply to a small region of spacetime, within which Eqs. (2.2) and (2.37) apply to good accuracy. A familiar example of a locally inertial frame in this sense is provided by a space station. Gravity is abolished to good accuracy within the space station, but the abolition is not perfect because the lines of gravitational force within the space station are not perfectly parallel. Also, gravity is slightly stronger on the Earth side of the space station than on the opposite side. As a result the effect of gravity can be felt within the space station, albeit at a very low level.

2.5 CURVATURE TENSOR

If we could choose Minkowski coordinates, the covariant derivative would become an ordinary derivative. Then, acting on any tensor, the commutator $[D_\mu, D_\nu]$ would vanish. In reality there are no Minkowski coordinates and the commutator does not

---

[7]To be precise, this is one of the meanings attached to the term "equivalence principle". The term is also used with somewhat different, though closely related, meanings.

[8]For quantum field theory we deal also with spinors but they behave like 4-vectors and tensors as far as the present discussion is concerned.

vanish. Acting on a 4-vector it gives

$$[D_\alpha, D_\beta]A^\mu = R^\mu{}_{\nu\beta\alpha}A^\nu, \tag{2.41}$$

where

$$R^\mu{}_{\nu\beta\alpha} = \frac{\partial\Gamma^\mu_{\nu\beta}}{\partial x^\alpha} - \frac{\partial\Gamma^\mu_{\nu\alpha}}{\partial x^\beta} + \Gamma^\mu_{\sigma\alpha}\Gamma^\sigma_{\nu\beta} - \Gamma^\mu_{\sigma\beta}\Gamma^\sigma_{\nu\alpha}. \tag{2.42}$$

The object $R^\mu{}_{\nu\beta\alpha}$ is called the curvature tensor. Using Eqs. (2.19) and (2.6) one can verify that $R^\mu{}_{\nu\beta\alpha}$ is indeed a tensor.[9] The curvature tensor vanishes if there are Minkowski coordinates and one can show that the converse is also true. It is therefore possible to find Minkowski coordinates if and only if the curvature tensor vanishes throughout spacetime.

From the curvature tensor one can form the Ricci tensor,

$$R_{\mu\nu} = R^\lambda{}_{\mu\lambda\nu}, \tag{2.43}$$

and the curvature scalar

$$R = R^\mu{}_\mu \tag{2.44}$$

The curvature tensor is so called because of an analogy involving curved surfaces. In the presence of the curvature tensor it is impossible to find Minkowski coordinates, which is analogous to the fact that it is generally impossible to find Cartesian coordinates on a curved surface.[10] Because of this analogy one says that spacetime is flat if the curvature tensor vanishes and curved if it does not. According to this terminology, gravity causes spacetime to be curved.

2.6 GEODESICS

Now I consider the worldline of an object subject to no force except gravity, which is called a spacetime geodesic. According to Special Relativity Newton's first law holds, which states that the object moves with constant velocity. That corresponds, for the trajectory $x^i(t)$ $(i = 1, 2, 3)$, to

$$\frac{d}{dt}\frac{dx^i}{dt} = 0. \tag{2.45}$$

Since $x^0 = t$, this can also be written

$$\frac{d}{dt}\frac{dx^\mu}{dt} = 0. \tag{2.46}$$

Taking gravity into account, the equivalence principle requires that this applies at each point on the trajectory in some locally inertial frame.

The spacetime geodesic is so called because of the analogy with a geodesic on

---

[9]This also follows from what is called the quotient theorem, which states that a quantity is a tensor if it yields a tensor when contracted with an arbitrary tensor, which in this case is A^ν.

[10]The only exceptions are the cylinder and the cone.

the surface of the Earth. Within an infinitesimal region around each point on that geodesic, we can set up Cartesian coordinates $\{x, y\}$, and the geodesic satisfies $d^2x/d\ell^2 = d^2y/d\ell^2$ where ℓ is the distance along the geodesic.

Transforming at each point to a generic coordinates system, Eq. (2.46) becomes

$$\frac{D}{D\lambda}\left(\frac{dx^\mu}{d\lambda}\right) = 0, \tag{2.47}$$

where λ is any parameter such that $d\lambda = dt$ in some locally inertial frame at each point along the trajectory. Such a parameter is called an affine parameter. Using Eq. (2.24) this gives

$$\frac{d^2x^\mu}{d\lambda^2} = -\Gamma^\mu_{\alpha\beta}\frac{dx^\alpha}{d\lambda}\frac{dx^\beta}{d\lambda}. \tag{2.48}$$

If we are dealing with an object that has non-zero mass, we can take λ to be the proper time (the one measured by a clock moving with the object) so that we have

$$\frac{d^2x^\mu}{d\tau^2} = -\Gamma^\mu_{\alpha\beta}\frac{dx^\alpha}{d\tau}\frac{dx^\beta}{d\tau}. \tag{2.49}$$

If the object is moving with speed $v \ll 1$, we have to a good approximation $v^i = dx^i/dt$ and $dt/d\tau = 1$. Then Eq. (2.49) gives[11]

$$\frac{dv^i}{dt} = -\Gamma^i_{00}. \tag{2.50}$$

In all of this, the object is supposed to have negligible size, and a negligible amount of rotation. If an object has negligible size but significant rotation, the object has acceleration in a locally inertial frame, which depends on the amount and direction of the rotation and on the spacetime curvature. This acceleration is also felt presumably by massive particles (elementary and composite). Therefore an electron, say, will be accelerating and emitting electromagnetic radiation at some level, even in the absence of an electromagnetic field. No calculation seems to exist concerning this effect for an astronomical source, but it has not been observed and one guesses will never be observed. The worldline of a photon in a locally inertial frame exactly satisfies Eq. (2.46) which means that it travels exactly on a geodesic.

EXERCISES

1. Taking an object to have speed V in some inertial frame, use the Lorentz transformation to calculate its speed in an inertial frame that moves with speed v in the opposite direction to the object. Compare this relativistic velocity addition formula with the non-relativistic velocity addition formula.

---

[11]This equation assumes that the spatial components Γ^i_{jk} are not exceptionally large. I will invoke it for Newtonian gravity and for the motion of a particle in a gravitational wave, the assumption being valid in both cases.

2. Suppose that an astronaut travels in a straight line from the Earth with constant speed $10^{-3}c$, and after ten years of Earth time turns around to return to the Earth at the same speed. Use the Lorentz transformation to calculate the amount by which the astronaut has aged. What assumptions did you make in arriving at your estimate? (This has been called the twin paradox, though it is not actually in conflict with Special Relativity.)

3. Show that $g_{\mu\nu}g^{\mu\nu} = 4$.

4. Verify Eq. (2.4), using Eq. (2.3).

5. Verify Eq. (2.8) using Eq. (2.6).

6. Insert Eq. (2.7) into Eq. (2.12), to verify Eq. (2.13).

7. Use Eq. (2.4) to find the components of the metric tensor in flat spacetime with spherical polar spatial coordinates.

8. Calculate $\Gamma^{\lambda}_{\mu\nu}$ for flat spacetime with spherical polar space coordinates. Verify that the curvature tensor given by Eq. (2.42) vanishes. In the geodesic equation (2.48), take the motion to be in the plane $\theta = \pi/2$ and write it down in spherical polar coordinates. Verify that it corresponds to motion in a straight line.

9. Verify Eq. (2.42).

10. Show that Eq. (2.40) corresponds to 40 different equations.

11. Propose an experiment that can be done within a space station, to verify that gravity is slightly stronger on the side nearest the Earth than on the opposite side.

CHAPTER **3**

General Relativity

CONTENTS

In the previous chapter we saw how to formulate the laws of physics, given the spacetime metric. In this chapter I describe General Relativity without assuming a previous knowledge of the subject. I continue to set $c = 1$.

3.1 ENERGY-MOMENTUM TENSOR

According to General Relativity, gravity is caused by a symmetric tensor $T^{\mu\nu}$, called the energy-momentum tensor. The energy-momentum tensor varies smoothly with position.

I will first define $T^{\mu\nu}$ in flat spacetime. At a given spacetime point, $T^{00} = T_{00}$ is the energy density which I will denote by ρ, and $T^{0i} = -T_{0i}$ is the momentum density. Taken together they are the density of four-momentum. The rest frame of the fluid is the one in which $T^{0i} = 0$, and in that frame the space components $T^{ij} = T_{ij}$ define the stress tensor. If it is isotropic then T_{ij} is the unit matrix times the pressure P. Otherwise we can write

$$T_{ij} = P\delta_{ij} + \Sigma_{ij}, \tag{3.1}$$

where the traceless matrix Σ_{ij} specifies the anisotropic part of the stress.

Let us suppose that Eq. (3.1) applies in the rest frame, and make a Lorentz boost to another frame with relative velocity $\mathbf{v}$ which has speed $v \ll 1$. Using the fact that $T^{\mu\nu}$ transforms like a product of 4-vectors, one can show that to first order in v the boost gives

$$T^{0i} = v^i \left(\rho + P \right), \tag{3.2}$$

with the other components unchanged.

For a gas, kinetic theory gives

$$P = \frac{1}{3}\rho \,\overline{v^2} \tag{3.3}$$

where $\overline{v^2}$ is the mean-square speed of the constituents. This applies to the total, and also to each constituent. For an ordinary gas we have $\overline{v^2} \ll 1$ which gives $P \ll \rho$. The latter is true also for solids and liquids. In these cases, Eq. (3.2) gives the usual result that the components of the momentum density are v^i times the mass density. In contrast, for photons in a gas we have $\overline{v^2} = 1$ giving $P = \rho/3$, and that is true to a good approximation for neutrinos in a gas. As we will see, the early Universe is gaseous and in that context species with $P \ll \rho$ are called matter, and species with $P = \rho/3$ (exactly or to a good approximation) are called radiation. Equivalently, matter species have $\overline{v^2} \ll 1$ and radiation species have (exactly or to a good approximation) $\overline{v^2} = 1$.

The energy-momentum tensor satisfies the continuity equation

$$\partial T^{\mu\nu}/\partial x^\mu = 0, \tag{3.4}$$

i.e.

$$\frac{\partial T^{0\nu}}{\partial t} = -\frac{\partial T^{j\nu}}{\partial x^j}. \tag{3.5}$$

I will call the $\nu = 0$ component the energy continuity equation, and I will call the $\nu = 1, 2, 3$ components the acceleration equation because they give the rate of change of the momentum density. (In cosmology literature the latter are often called the Euler equation.)

Integrating over a region of space using Minkowski coordinates, one finds

$$\frac{\partial}{\partial t}\int dV T^{0\nu} = -\int dS_j \frac{\partial T^{j\nu}}{\partial x^j}, \tag{3.6}$$

where the right hand side goes over the boundary of the region. The left hand side is the rate of change of the four-momentum within the region. For a surface surrounding an isolated system, the right hand side vanishes. This shows that the four-momentum of an isolated system is constant.

In curved spacetime, the previous paragraph does not apply because there are no Minkowski coordinates, but the preceding paragraphs apply at each point in a locally inertial frame. Going to generic coordinates, Eq. (3.4) becomes

$$D_\mu T^{\mu\nu} = 0., \tag{3.7}$$

i.e.

$$\frac{\partial T^{\mu\nu}}{\partial x^\mu} + \Gamma^\mu_{\mu\alpha} T^{\alpha\nu} + \Gamma^\nu_{\mu\alpha} T^{\alpha\mu} = 0.. \tag{3.8}$$

3.2 EINSTEIN FIELD EQUATION

According to General Relativity, the energy-momentum tensor is related to the curvature tensor by Einstein's field equation which reads

$$R_{\mu\nu} - \frac{1}{2}g_{\mu\nu}R = 8\pi G T_{\mu\nu}. \tag{3.9}$$

The field equation is a fundamental equation of physics, like Maxwell's equations. It is a second-order partial differential equation for the metric $g_{\mu\nu}$, which shows that the sources of gravity according to General Relativity are the energy density, momentum density and stress. In contrast, Newtonian gravity has only a single source of gravity which is the mass density.

The form of the left hand side is dictated by the requirement that it leads to a second-order partial differential equation for $g_{\mu\nu}$ and that it is consistent with the continuity equation $D_\mu T^{\mu\nu} = 0$. The consistency requires

$$D_\mu \left(R^{\mu\nu} - \frac{1}{2}g^{\mu\nu}R \right) = 0. \tag{3.10}$$

To check that this holds, we should go to a locally inertial frame in which D_μ becomes $\partial/\partial x^\mu$, which I will denote by ∂_μ. In such a frame, only the first two terms of Eq. (2.42) survive, and using Eq. (2.14) they give

$$D_\lambda R^\mu{}_{\nu\beta\alpha} = D_\lambda \partial_\alpha \Gamma^\mu_{\nu\beta} - D_\lambda \partial_\beta \Gamma^\mu_{\nu\alpha}, \tag{3.11}$$

which implies

$$D_\lambda R^\mu_{\nu\alpha\beta} + D_\beta R^\mu_{\nu\lambda\alpha} + D_\alpha R^\mu_{\nu\beta\lambda} = 0. \tag{3.12}$$

These are called the Bianchi identities. Since the covariant derivative of the metric tensor vanishes, the Bianchi identities still hold if we raise λ. Contracting that index with ν we obtain

$$D_\eta R_{\mu\kappa} - D_\kappa R_{\mu\eta} + D_\nu R^\nu{}_{\mu\kappa\eta} = 0. \tag{3.13}$$

Contracting again gives

$$D_\eta R - D_\mu R^\mu{}_\eta - D_\nu R^\nu{}_\eta = 0. \tag{3.14}$$

Since the covariant derivative of the metric tensor vanishes, this is equivalent to Eq. (3.10).

3.3 SCHWARZSCHILD METRIC AND BLACK HOLES

Using the angular spherical polar coordinates θ and ϕ, the distance $d\ell$ between nearby points on a sphere is given by

$$d\ell^2 = r^2 \left(d\theta^2 + r^2 d\theta^2 \right), \tag{3.15}$$

where r is a constant. Integration over the surface gives its area as $4\pi r^2$, but in curved spacetime its radius may not be r.

Outside a spherically symmetric object, the line element has the form

$$ds^2 = -\left(1 - \frac{r_s}{r}\right)dt^2 + \left(1 - \frac{r_s}{r}\right)^{-1}dr^2 + r^2\left(d\theta^2 + \sin^2\theta\, d\varphi^2\right), \qquad (3.16)$$

where $r_s = 2GM$. Indeed, this metric is the only spherically symmetric solution of the Einstein field equation with $T_{\mu\nu} = 0$, that reduces to flat spacetime in the limit $r \to \infty$. It is called the Schwarzschild metric, and r_s is called the Schwarzschild radius. We see that with this metric, the radius of a sphere with area $4\pi r^2$ is *not* equal to r because that would require the coefficient of dr^2 to be 1 (both outside and inside the object).

Since the metric components are independent of t, we can construct from any trajectory $x^i(t)$ another trajectory $x^i(t + \Delta t)$. In other words, objects following the same path are separated by some constant Δt in the time coordinate. Their separation in time measured by stationary clocks on the other hand, is $(1 - r_s/r)^{1/2}$ which varies with height. A clock at a lower position therefore experiences time dilation, compared with a clock at a higher position.

At the surface of an ordinary object, $r \ll r_s$. Still, the time dilation here on Earth is big enough to measure in the laboratory using the Mössbauer effect. The GPS system, used to locate position on the Earth, employs signals from about 30 satellites and it has to take into account time dilation and other General Relativistic effects. So it is General Relativity, not Special Relativity, that has finally become of practical significance!

Using the geodesic equation with the Schwarzschild metric one can work out the deflection of a light ray as it passes by the edge of the sun. Working in the limit $r/r_s \to 0$ which is an excellent approximation, one finds that the angle of deflection is $4GM/c^2R$ where M and R are the mass and radius of the sun. This result, obtained from General Relativity, is exactly twice the result that one would get by applying Newton's theory of gravity to the motion of a photon. The discrepancy is not surprising, because Newton's theory applies only to an object moving with speed much less than c. In 1919 Eddington measured the angle by observing a star near the edge of the sun at a solar eclipse, and announced that the result confirmed the prediction of General Relativity. This was reported in newspapers all over the world, placing Einstein firmly in the public consciousness, though there is controversy as to whether the accuracy of Eddington's measurements was sufficient to justify the announcement.

A non-rotating black hole is defined as an object for which the Schwarzschild metric applies all the way down to $r = r_s$. Outside the black hole, a photon moving radially outwards has

$$\frac{dr}{dt} = \left(1 - \frac{r_s}{r}\right), \qquad (3.17)$$

because that corresponds to $ds^2 = 0$, and a massive object moving radially outwards has smaller dr/dt corresponding to $ds^2 > 0$. If $r = r_s$ the trajectory of a photon moving radially outwards satisfies $dr/dt = 0$, which means that neither light nor anything else can emerge from the black hole. The sphere $r = r_s$ may be regarded as the surface of the black hole. More generally a black hole, whether rotating or not, is defined as an object from which light cannot escape. The most general black hole is defined by its mass and its angular momentum. Its metric is called the Kerr metric, which I will not consider here.

3.4 WEAK GRAVITY

In most situations, gravity is weak corresponding to the metric

$$g_{\mu\nu} = \eta_{\mu\nu} + h_{\mu\nu}, \tag{3.18}$$

with $|h_{\mu\nu}| \ll 1$. Since $h_{\mu\nu}$ is small it can be treated to first order.

Let us see how $h_{\mu\nu}$ changes if we go from some coordinates x^μ to new ones. To keep $h_{\mu\nu}$ small, the change in the coordinates should be small. A choice of coordinates that keeps $h_{\mu\nu}$ small is called a gauge, and the transformation from one gauge to another is called a gauge transformation. Let us denote the small change by ξ^μ;

$$x'^\mu = x^\mu + \xi^\mu \tag{3.19}$$

$$\frac{\partial x'^\mu}{\partial x^\nu} = \delta^\mu_\nu + \frac{\partial \xi^\mu}{\partial x^\nu}. \tag{3.20}$$

To keep $h_{\mu\nu}$ small we need $|\partial \xi^\mu / \partial x^\nu| \ll 1$, and working to first order in that quantity Eq. (2.11) gives the gauge transformation

$$h'_{\mu\nu} = h_{\mu\nu} - \frac{\partial \xi_\mu}{\partial x^\nu} - \frac{\partial \xi_\nu}{\partial x^\mu}, \tag{3.21}$$

where $\xi_\mu \equiv \eta_{\mu\nu} \xi^\nu$.

It is convenient to regard $h_{\mu\nu}$ as a field, like the electromagnetic field, that lives in flat spacetime. Then one can raise one or both of its indices by contracting with $\eta^{\mu\nu}$ and it is useful to consider

$$\bar{h}_{\mu\nu} \equiv h_{\mu\nu} - \frac{1}{2}\eta_{\mu\nu}h, \tag{3.22}$$

where $h \equiv h^\mu_{\ \mu}$ is the trace of $h^\mu_{\ \nu}$.

Since $h_{\mu\nu}$ is supposed to live in flat spacetime, the gauge transformation corresponds simply to a change in the field $h_{\mu\nu}$, not to a change in the coordinates. I will now show that it is possible to choose the gauge so that $\partial \bar{h}^{\mu\nu}/\partial x^\mu = 0$. To do that I need the gauge transformation of $\bar{h}^{\mu\nu}$. Using Eqs. (3.21) and (3.22) one finds

$$\bar{h}'_{\mu\nu} = \bar{h}_{\mu\nu} - \frac{\partial \xi_\mu}{\partial x^\nu} - \frac{\partial \xi_\nu}{\partial x^\mu} + \eta_{\mu\nu}\frac{\partial \xi^\alpha}{\partial x^\alpha}. \tag{3.23}$$

We can make $\partial \bar{h}'^{\mu\nu}/\partial x^\mu = 0$ by requiring

$$\Box \xi^\mu = \frac{\partial \bar{h}^{\mu\nu}}{\partial x^\nu}. \tag{3.24}$$

From now on I assume $\partial \bar{h}^{\mu\nu}/\partial x^\mu = 0$. This requirement does not determine $\bar{h}^{\mu\nu}$ completely, because it is still satisfied if we make a further transformation using a ξ^μ for which

$$\Box \xi^\mu = 0. \tag{3.25}$$

Evaluating the Einstein field equation to first order, one finds

$$\Box \bar{h}_{\mu\nu} = -16\pi G T_{\mu\nu}. \tag{3.26}$$

This simple form is the reason for considering $\bar{h}_{\mu\nu}$.

3.5 NEWTONIAN GRAVITY

Newton's theory of gravity assumes the existence of a universal time and Cartesian coordinates. The acceleration of a particle with trajectory $\mathbf{r}(t)$ is given by

$$\frac{d^2\mathbf{r}}{dt^2} = -\boldsymbol{\nabla}\phi(\mathbf{r}, t). \tag{3.27}$$

The gravitational potential ϕ is related to the mass density ρ by the Poisson equation

$$\nabla^2 \phi(\mathbf{r}, t) = 4\pi G \rho(\mathbf{r}, t). \tag{3.28}$$

The solution of the Poisson equation which vanishes at infinity is

$$\phi(\mathbf{r}, t) = -G \int d^3 r' \frac{\rho(\mathbf{r}', t)}{|\mathbf{r} - \mathbf{r}'|}. \tag{3.29}$$

A small isolated object with mass M at position $\mathbf{r} = 0$ gives the potential

$$\phi(r) = -\frac{GM}{r}. \tag{3.30}$$

Putting that into Eq. (3.27) gives the inverse square law for the acceleration of an object with position $\mathbf{r}(t)$:

$$\frac{d^2\mathbf{r}}{dt^2} = -GM\frac{\mathbf{r}}{r^3}. \tag{3.31}$$

The mass continuity equation gives the time dependence of ρ, within a fluid element which is moving with velocity $\mathbf{u}$. It is

$$\frac{d\rho(\mathbf{x}, t)}{dt} = -(\boldsymbol{\nabla} \cdot \mathbf{u})\,\rho(\mathbf{x}, t), \qquad \frac{d}{dt} \equiv \frac{\partial}{\partial t} + \mathbf{u} \cdot \boldsymbol{\nabla}. \tag{3.32}$$

This equation says that the change in ρ is caused by the movement of mass in or out of the element. Applying Newton's equation $\mathbf{F} = m\mathbf{a}$ to the fluid element gives its acceleration

$$\frac{d\mathbf{u}}{dt} = -\frac{1}{\rho}\boldsymbol{\nabla}P - \boldsymbol{\nabla}\phi.. \tag{3.33}$$

Let us see what is needed to obtain Newtonian gravity as an approximation to General Relativity, setting $c = 1$. We need $|T^{00}| \gg |T^{01}| \gg |T^{ij}|$. (In the case of a gas these inequalities just correspond to the constituents moving with speed $\ll 1$.) We need the time derivative of $T^{00} = \rho$ to be much less than its gradient. We also need $|\bar{h}^{00}| \gg |\bar{h}^{0i}| \gg |\bar{h}^{ij}|$ with $\bar{h}^{00}$ slowly varying, which makes Eq. (3.26) consistent. Then it is a good approximation to replace $\Box$ by ∇ on the left hand side of Eq. (3.26). The 00 component of Eq. (3.26) is then

$$\nabla^2 \bar{h}^{00} = -16\pi G \rho. \tag{3.34}$$

This corresponds to the Poisson equation of Newtonian gravity if we make the identification

$$\bar{h}^{00} = -4\phi. \tag{3.35}$$

Since $\bar{h}^{00}$ is the dominant contribution to $\bar{h}^{\mu\nu}$ we have to a good approximation $h = \bar{h}^{00}$. Then, to have $|\bar{h}^{ij}| \ll |\bar{h}^{0i}| \ll |\bar{h}^{00}|$, Eq. (3.22) requires to good accuracy

$$h^{00} = -2\phi \tag{3.36}$$
$$h^{i0} = 0 \tag{3.37}$$
$$h^{ij} = \frac{1}{2}\delta^{ij} h = \frac{1}{2}\delta^{ij}\bar{h}^{00} = 2\delta^{ij}\phi, \tag{3.38}$$

giving the line element

$$ds^2 = -(1 + 2\phi)\,dt^2 + (1 - 2\phi)\left(dx^2 + dy^2 + dz^2\right). \tag{3.39}$$

Newtonian gravity assumes a universal time and Cartesian coordinates, which would require $\phi = 0$. That gives a good approximation to Eq. (3.39) provided that $|\phi| \ll 1$, which we therefore need for the validity of Newtonian gravity. Finally, evaluating Γ^i_{00} to first order in ϕ, the geodesic equation (2.50) reproduces Eq. (3.27).

3.6   GRAVITATIONAL WAVES

Gravitational waves are the oscillating solutions of Eq. (3.26). For a gravitational wave travelling through the vacuum, the right hand side of this equation vanishes, and there is a plane wave in the vacuum of the form

$$\bar{h}_{\mu\nu} = \mathrm{Re}\left[A_{\mu\nu}e^{ik_\alpha x^\alpha}\right] = \mathrm{Re}\left[A_{\mu\nu}e^{i(\mathbf{k}\cdot\mathbf{r} - \omega t)}\right], \tag{3.40}$$

with $\omega = k$. The condition $\partial\bar{h}_{\mu\nu}/\partial x^\mu = 0$ corresponds to $k^\mu A_{\mu\nu} = 0$. Having satisfied this requirement, we can still make a further gauge transformation using a ξ^μ which satisfies Eq. (3.25). For a plane wave it has to be of the form

$$\xi^\mu = \epsilon^\mu e^{ik_\alpha x^\alpha}, \tag{3.41}$$

with ϵ^μ arbitrary. Going to lower-index quantities this gives

$$A'_{\mu\nu} = A_{\mu\nu} - i\epsilon_\mu k_\nu - i\epsilon_\nu k_\mu + i\eta_{\mu\nu}\epsilon^\alpha k_\alpha. \tag{3.42}$$

Let us choose the z axis as the **k** direction. Then it is not difficult to show that we can choose ϵ^μ so that $\bar{h}^{0i} = 0$ and $h = 0$. This is called a transverse traceless or TT gauge. Since $h = 0$ there is no difference between $\bar{h}_{\mu\nu}$ and $h_{\mu\nu}$ so that we can revert to the latter. Let us define $h_{+,\times}$ by

$$h_{\mu\nu} = \mathrm{Re}\left[\left(e^+_{\mu\nu}h_+(\mathbf{k}) + e^\times_{\mu\nu}h_\times(\mathbf{k})\right)e^{i(\mathbf{k}\cdot\mathbf{r}-\omega t)}\right]. \tag{3.43}$$

Fixing on a particular choice for the orientation of the x and y axes we define the nonzero components of $e^+_{\mu\nu}$ and $e^\times_{\mu\nu}$

$$e^\times_{xy} = e^\times_{yx} = e^+_{xx} = -e^+_{yy} = 1. \tag{3.44}$$

The components for any other coordinate choice are given by the tensor transformation law.

To see how gravitational waves might be detected, let us look at the effect of a gravitational plane wave on a free particle. To first order in $\bar{h}_{\mu\nu}$, Eq. (2.19) gives $\Gamma^i_{00} = 0$ and then Eq. (2.50) gives $d^2x^i/dt^2 = 0$. Therefore, if dx^i/dt vanishes at some instant it vanishes forever making x^i constant. We need only consider that case, because a non-vanishing dx^i/dt would just mean that the particle is drifting with respect to the gravitational wave. We are interested in the effect on a particle that is not drifting.

Taking x^i to be constant then, let us consider the effects of h_+ and $h_\times$. Starting with h_+, I consider a real plane wave travelling in the **k** direction, obtained from Eq. (3.40). Taking h_+ to be real we can write

$$h_{\mu\nu} = e^+_{\mu\nu}h_+(\mathbf{k})\sin(\mathbf{k}\cdot\mathbf{x} - \omega t). \tag{3.45}$$

Taking **k** as the z direction, the distance in the transverse direction is given by

$$d\ell^2 = [1 + h_+\sin(kz - \omega t)]\,dx^2 + [1 - h_+\sin(kz - \omega t)]\,dy^2. \tag{3.46}$$

The distances in the x and y directions are therefore (working as always to first order in $h_{\mu\nu}$)

$$\ell_x = \left(1 + \frac{1}{2}h_+\sin(kz - \omega t)\right)x \tag{3.47}$$

$$\ell_y = \left(1 - \frac{1}{2}h_+\sin(kz - \omega t)\right)y \tag{3.48}$$

We see that the distance is stretching in the x direction while contracting in the y direction, and vice versa. The effect on a ring of particles is shown in Figures 3.6 and 3.6.

The effect of $h_\times$ is the same as that of h_+ except for rotation by $45°$. To see this, start with an analogue of Eq. (3.46);

$$d\ell^2 = dx^2 + dy^2 + 2h_\times\sin(kz - \omega t)dxdy \tag{3.49}$$

Rotate by 45° we get coordinates x' and y' given by

$$x = \frac{1}{\sqrt{2}}(x' + y') \tag{3.50}$$

$$y = \frac{1}{\sqrt{2}}(x' - y'), \tag{3.51}$$

which gives

$$d\ell^2 = dx'^2 + dy'^2 + h_\times \sin(kz - \omega t)\left(dx'^2 - dy'^2\right). \tag{3.52}$$

This is indeed the same as Eq. (3.46), with $h_\times$ instead of h_+. The effect of $h_\times$ on a ring of free particles is shown in Figure 3.6.

If a particle is fixed, the acceleration a that would be experienced by a free particle is converted into a force $F = ma$ where m is the mass of the particle. This allows one to calculate the stress that a gravitational wave causes, when it passes through a solid object. The measurement of this stress is one strategy for detecting gravitational waves. The other strategy is to detect the wave's effect on distance, as described in the previous paragraphs.

Going to the quantum theory, a gravitational wave corresponds to a beam of gravitons, the graviton being a particle species. Rotating the state vector of a graviton by ϕ radians about its direction of motion has the effect

$$|\rangle \to e^{i\hat{J}_z\phi}|\rangle = e^{2i\phi}|\rangle, \tag{3.53}$$

where $\hat{J}_z$ is the operator corresponding to the component of angular momentum along the z axis, taken to be the direction of motion. This means that the graviton has spin 2. The factor 2 means that the state is unchanged if we rotate by 180°, which is also clear from our discussion of gravitational waves. This may be compared with the case of an electromagnetic wave, which is unchanged only if we rotate by 360°. As a result, the 2 in Eq. (3.53) becomes 1 and the photon has spin 1.

Like an electromagnetic wave, a gravitational wave can be generated in two ways. First, it is generated at some level by any object that is accelerating. This is like the generation of an electromagnetic wave by any *charged* object that is accelerating, and we will see how it works in Chapter 4. Second, individual gravitons can be created just as individual photons can be created. We will see how that may work in Chapter 18.

Figure 3.1 The movement of a ring of free particles, caused by h_+ for a gravitational wave travelling in the orthogonal direction.

Figure 3.2 The movement of a ring of free particles, caused by $h_\times$.

EXERCISES

1. Write down the expression which gives the trace of a 3×3 matrix in terms of its diagonal components, and hence verify that Eq. (3.1) is valid for a unique choice of P and Σ_{ij}.

2. Verify the statement made in the sentence containing Eq. (3.2).

3. Use Eq. (3.16) and the condition $ds^2 = 0$ to show that a photon travelling radially outwards on the surface of a black hole satisfies $dr/dt = 0$. Why does this prove (more simply) that nothing can emerge from a black hole?

4. Calculate the Schwarzschild radius of a black hole whose mass is that of the sun ($M_\odot = 2.0 \times 10^{30}$ kg). Then apply the appropriate scaling to write down the Schwarzschild radius of a black hole with mass $10^6 M_\odot$. (As seen in Chapter 4, these are very roughly the masses of the two types of black hole that are known to exist in the Universe.)

5. The Earth's mass is 5.97×10^{24} kg and its radius is 6.37×10^6 m. Using this information, estimate the fraction by which a clock on the floor of your room runs more slowly than a clock on the ceiling.

6. Verify the gauge transformations (3.21) and (3.23).

7. Calculate to first order the Levi-Civita connection and the curvature tensor, generated by $\bar{h}_{\mu\nu}$. Hence verify the field equation Eq. (3.26).

8. Verify the line element (3.39).

9. Verify the statement about e^μ made after Eq. (3.42).

II

The Big Bang

The present Universe

CONTENTS

In this chapter I will describe the Universe at the present epoch. My account has to begin with a disclaimer. We don't know about the whole Universe. Instead, we know only about the part of the Universe that we can see — the observable Universe. The boundary of the observable Universe is a sphere around us, called the horizon.[1] This limitation on our knowledge is not as serious as one might think though, because the observable Universe is almost homogeneous. To be precise, the properties of a region containing many galaxies are more or less independent of the location of that region. It is reasonable to expect that the Universe remains homogeneous as we go beyond the horizon, for at least some distance. If that expectation is satisfied, you can take 'Universe' to mean, not just the observable Universe but the entire homogeneous patch around us. In this matter, we are in a position similar to a sailor at sea. The sailor may reasonably expect that the sea extends beyond the horizon in all directions, but that could be wrong.

To describe the present Universe we need a unit of mass and a unit of distance. For mass one generally uses the mass of the sun,

$$M_{\odot} = 1.99 \times 10^{30}\,\text{kg}. \tag{4.1}$$

For distance, the basic unit for astronomy is the parsec. This is the distance at which the diameter of the Earth's orbit subtends and angle of 1 arcsecond. In practice one generally uses the Mpc, equal to 10^6 parsecs. It is related to more familiar units by

$$1\,\text{pc} = 30.86 \times 10^{15}\,\text{m} = 3.262\,\text{light-years}. \tag{4.2}$$

As we will see in Chapter 8, the distance to the horizon is about 14 Gpc.

4.1 STARS AND GALAXIES

In this section I consider stars, galaxies and galaxy clusters, whose sizes and masses are listed in Table 4.1. A typical star is similar to the sun, which is a ball of

---

[1] I will give a more objective definition of the horizon later.

hot gas consisting of freely moving nuclei and electrons. The radiation from the star comes from nuclear fusion, which creates both neutrinos and photons. The neutrinos escape promptly without undergoing significant interactions. The photons by contrast bounce off the electrons and it takes about a million years for a photon to escape from the centre of the sun. Some stars are known to have planets circling round them. It is still a matter of speculation whether life may have developed on one of those planets.

When a star has used up all of its nuclear fuel, it collapses under its own weight. The collapse of a more or less spherical ball of gas is a common occurrence in the Universe. The collapse generates random motion of the particles, whose kinetic energy is drawn from the gravitational potential energy. It is halted if and when the average kinetic energy is about equal to the gravitational potential energy.

If the star is not too heavy, the collapse of the star is halted by what is called electron degeneracy pressure. This refers to the fact that as the collapse proceeds, the quantum states with low electron momentum become fully occupied which fixes their contribution to the pressure. When the collapse is halted by electron degeneracy, we have a white dwarf consisting of highly compressed ordinary matter.

The mass of a white dwarf cannot be more than about $1.44M_\odot$, because electron degeneracy cannot resist the pull of gravity for a heavier star. A heavier star will become so highly compressed that each proton combines with an electron, to form a neutron and a neutrino. The neutrinos fly away but the neutrons remain, and for all except the very heaviest stars, the collapse is finally halted by neutron degeneracy pressure which works in precisely the same way as electron degeneracy pressure. We then have a neutron star consisting of neutrons. The radius of a neutron star is only around 10 km, and it is so dense that a teaspoon of a neutron star is about a thousand times as heavy as the Great Pyramid of Giza. Some neutron stars are emitting a strong beam of electromagnetic radiation which rotates like a lighthouse beam so that we see it as a series of pulses. Those neutron stars are called pulsars.

A neutron star cannot be more than about three times heavier than the sun, because even neutron degeneracy cannot resist the pull of gravity for a heavier star. A heavier star will keep on collapsing, to form a black hole with mass roughly $M_\odot$.

The violent collapse leading to the formation of a neutron star or black hole is called a supernova. A supernova first emits an intense burst of neutrinos, lasting for less than a minute. Then there is an intense burst of photons lasting a few days. If a supernova occurs in our galaxy, and is not obscured by dust, it can outshine all other stars and be visible during the day. Those naked-eye supernovas are very rare and only a few have been described by past generations, the last one in 1604.

Supernovas are thought to account for what are called gamma ray bursts, which are observed at the rate of around one a day. The burst of gamma rays is apparently released as two oppositely directed jets, just after a star collapses to form a black hole. The amount of energy released is huge, around one percent of the star's rest mass.

Supernovas are the most violent events in the Universe. They are vital to our existence, and even to the existence of the Earth itself. That is because they are responsible for the creation of practically all species of nuclei. Without supernovas, the present Universe would contain only the hydrogen and helium, plus tiny amounts of the lithium and beryllium. (We will see how *those* originate in the Chapter 5.)

Many stars are in pairs that orbit around each other. These are called binaries. If a binary contains a white dwarf and an ordinary star, matter can flow from the star to the white dwarf. If the mass of the white dwarf becomes bigger than $1.44M_\odot$, it collapses to become a neutron star. The collapse is a supernova, just like the collapse of an isolated star.

The stars are gathered into **galaxies**, containing anything from about 10^6 to 10^9 stars. We are near the edge of a large galaxy, called the Milky Way galaxy. Looking at the sky with the naked eye, one sees stars in the Milky Way galaxy, and the Milky Way itself which consists of stars in our galaxy that can't be seen individually with the naked eye.[2]

At least the larger galaxies have a black hole at their centre. The mass of the black hole is typically around $10^6 M_\odot$. It accounts for only a tiny fraction of the galaxy's mass, but it is still far heavier than the black holes formed by the collapse of a star. Sometimes, a large amount of matter is falling violently into the black hole generating intense electromagnetic radiation. We then have what is called an active galactic nucleus.

Most galaxies are gathered into **galaxy clusters**, which can contain anything from just a few galaxies to around 10^3 galaxies. The galaxy clusters are evenly distributed throughout the observable Universe.

As summarised in Figure 4.1, gravitational waves are produced in a number of ways. They are produced by pairs of very nearby objects that are rotating around each other, called close binaries. Each object must be either a black hole or a neutron star, that being the only way in which they can be close enough to be a powerful source. The emission of the gravitational waves causes the distance between the pair to decrease so that their orbit spirals inwards. This has been observed in cases where one of the objects is a pulsar, and the rate of spiralling is in accordance with General Relativity.

The spiralling orbit of a close binary can lead eventually to the collision of the objects, leading to a final intense burst of gravitational waves. Such a burst, coming from a pair of colliding black holes, was observed in 2016 marking the first direct detection of gravitational waves and the beginning of gravitational wave astronomy. The observation was made with a laser interferometer in the USA, called LIGO.

Other sources of potential observable gravitational waves include (i) a white dwarf, neutron star or stellar black hole which is orbiting around a galactic black hole, (ii) a pair of galactic black holes orbiting around each other, (iii) a rotating neutron star and (iv) a supernova. Gravitational waves from a pair of galactic black

---

[2]The stars in the Milky Way were first seen by Galileo in 1610 using his telescope. Only three other galaxies are visible to the naked eye; these are the Andromeda Nebula seen in the northern hemisphere, and the Small and Large Magellanic Clouds seen in the southern hemisphere.

holes will be detected through their distortion of the otherwise regular pulses of the electromagnetic radiation observed from a pulsar.

4.2 PARTICLES IN THE UNIVERSE

The Universe contains protons, neutrons and electrons, which in the context of cosmology are collectively called baryons.[3] In addition, the Universe contains what is called Cold Dark Matter or CDM. The CDM is called Cold because its particles have hardly any random motion. It's called Dark because it doesn't emit any radiation. But it doesn't absorb any either, so it should really be called Cold Transparent Matter. Even more remarkably, there is no discernable effect when a CDM particle encounters a particle of baryonic matter, even though such encounters must be taking place all the time. As a result of all this, we know of the CDM only through its gravitational effect, and we don't know anything about the particles that it's made of. Some of the CDM is concentrated within galaxies and galaxy clusters, being held there by the force of gravity. Most of the CDM though, is pretty evenly distributed throughout the Universe.

The Universe contains electromagnetic radiation, in other words photons. Some of the photons were emitted by stars (and the gas between them) but most were emitted before there were any stars or galaxies, at an epoch called the epoch of last scattering. Before the epoch of last scattering the photons were bouncing off free electrons, but after it the electrons were bound into atoms which allowed the photons to travel freely. The photons emitted at last scattering are now seen as the Cosmic Microwave Background (CMB). The CMB is almost isotropic, reflecting the near-homogeneity of the early Universe, but its slight anisotropy is crucial for cosmology. As we shall see in Section 14.4 it tells us more about the early Universe than all other observations put together.

The Universe also contains neutrinos. Some of the neutrinos were emitted by stars, but most were emitted at what is called the epoch of neutrino decoupling. Before that epoch, neutrinos were constantly being created and annihilated, but after it each neutrino travelled freely. The neutrinos emitted at the epoch of neutrino decoupling constitute what one might call the Cosmic Neutrino Background. The Cosmic Neutrino Background has not been directly detected, but its existence can be inferred from the statistical properties of the CMB anisotropy.

The detection of neutrinos is difficult, because their interaction has such low probability that a typical neutrino passes right through the Earth without interacting. A neutrino detector consists of a large tank of fluid, with apparatus to observe the outcome of the rare neutrino collisions with baryonic matter. To shield it from cosmic rays, the fluid has to be placed deep underground or at the bottom of a large body of water. These detectors see neutrinos created at accelerators and neutrinos emitted by the sun. In 1987, three detectors saw neutrinos emitted by supernova

---

[3]Only the protons and neutrons are baryons in the terminology of particle physics, but they far outweigh the electrons which is the reason for including them under the heading of baryons in cosmology.

	Size	Mass
Sun	7×10^5 km	1
Star	10^4 km to 10^8 km	10^{-1} to 10^2
Stellar Black Hole	roughly 10 km	roughly 10
Galactic Black Hole	roughly 10^6 km	roughly 10^6
Galaxy	1 to 50 kpc	10^7 to 10^{13}
Galaxy cluster	2 to 10 Mpc	10^{14} to 10^{15}

Table 4.1 **Cosmic sizes and masses** The masses are in units of $M_\odot = 1.99 \times 10^{30}$ kg. The sun, stars, black holes and the observable Universe are all spherical objects and for them 'size' means 'radius'. Galaxies and galaxy clusters are at best only roughly spherical and the 'size' given for them is just a rough estimate of the radius of the sphere into which they would fit. The range of star sizes excludes neutron stars, whose radius is only about 10 km. The distance to the horizon is about 1.40×10^4 Mpc.

1987A. That detection was possible because the supernova occurred in a nearby dwarf galaxy, and such nearby supernovae are expected only two or three times per century. Starting in 2013, a neutrino background coming from unresolved sources is being observed by the IceCube detector under the Antarctic ice. This marks the beginning of neutrino astronomy.

Some objects in the Universe emit, in addition to photons and neutrinos, particles and nuclei that arrive at the Earth constituting what are called primary cosmic rays.[4] It's not known for sure what those objects are, but likely candidates include supernovas, quasars and active galactic nuclei. The same objects are presumably responsible for the very high energy gamma rays, that like the cosmic rays come to us from all directions rather than from identifiable sources.

Primary cosmic rays consist of electrons and nuclei (mostly hydrogen and helium) along with a tiny fraction of positrons and anti-protons. When they hit the atmosphere the primary cosmic rays produce more particles, called secondary cosmic rays. Cosmic rays are detected by a variety of methods.

---

[4]The particles and nuclei interact with the cosmic gas on their journey towards us. The products of such interactions, arriving at the Earth, are included under the heading 'primary cosmic rays'.

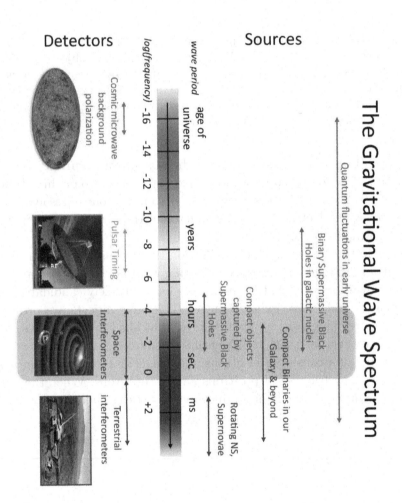

Figure 4.1   Gravitational radiation. Courtesy of NASA

EXERCISES

1. The first observation of gravitational waves, in 2016, detected an amplitude $|h_{+,\times}| \sim 10^{-22}$ and the waves came from colliding black holes with diameter very roughly 1 km which were roughly 10^9 light-years away from us. Using the fact that $h_{+,\times}$ is inversely proportional to the distance from the source (corresponding to an inverse square law for the intensity of the wave), estimate $|h_{+,\times}|$ when the waves first emerged, and explain why it is roughly what one would expect.

2. If two isolated protons are at rest and 10^{-10} m apart, each of them experiences an electric force of 2.3 Newtons. Using Eq. (3.31) and the values of G and the proton mass given in Appendix A.3, calculate the gravitational force between them. With p/e denoting the average number of protons per electron in the Universe, taken to be everywhere the same, calculate the value of $|p/e - 1|$ for which the gravitational and electric forces would cancel. What does that imply for the actual value of $|p/e - 1|$?

3. Using the data in the fifth line of Table 4.1, estimate the speed needed for a particle to emerge from a galaxy. What does that tell you about the random motion of the CDM?

4. Use the data in the last two lines of Table 4.1 to estimate the speed of an object in a circular orbit at the edge of (a) a galaxy and (b) a galaxy cluster. Explain why these are roughly the speeds of any object moving within the galaxy or galaxy cluster. Using that result, estimate the spread of redshifts that radiation from the galaxy or galaxy cluster will have, superimposed on the overall redshift caused by the motion of the object. (The observation of this spread, for radiation from a star or luminous gas, provides an estimate of the mass of the galaxy or galaxy cluster. By comparing it with the estimated mass of luminous matter, one arrives at an estimate of the mass of the dark matter.)

CHAPTER 5

A first look at the history

CONTENTS

5.1   THE BIG BANG AND INFLATION

We know, in essence, the history of the Universe from the time that it was about a second old, up to the present when it is 1.38×10^{10} yr old. We know it, because it is confirmed by observation to an extent that excludes beyond reasonable doubt any alternative. In contrast, the earlier history is a matter of conjecture.

To calculate the known history of the Universe, one needs the fundamental constants and some of the Standard Model parameters. One also needs some additional numbers, which I will call the fundamental cosmological parameters. These are displayed in Table A.3. The first five, which will be defined in this chapter, are needed to describe the homogeneous Universe.[1] The other two, which will be defined in Chapter 11, are needed to describe the slight inhomogeneity existing at early times.

Throughout the known history, the cosmic gas has been expanding and it presumably was expanding also before the known history. The entire era when the cosmic gas existed is called the **Big Bang**. The Big Bang was probably preceded by what is called **inflation**. During inflation, the Universe expanded at an ever-faster rate which was possible because it contained one or more fields instead of particles. Conditions at the end of a long era of inflation are more or less independent of what they were at the beginning. As a result, a suitably chosen inflation scenario can provide a complete explanation of what we observe. This is both a blessing and a curse. It's a curse because it means that observation can't distinguish between different scenarios for what happened before inflation. It's a blessing,

---

[1]The choice of these five fundamental parameters is not unique. One could to some extent exchange them with the derived parameters shown in Table A.4.

because it removes the need to think about such scenarios; for practical purposes, the history of the Universe begins with inflation!

Alternatives to inflation have been proposed, in which the very early Universe is contracting, the contraction only later giving way to the expansion that we now see. That's called a bouncing Universe. There are also 'cyclic' alternatives, in which contraction and expansion alternate. The alternatives are more complicated than inflation, and less popular.

5.2 EXPANSION OF THE UNIVERSE

To describe the expansion of the Universe, we can begin with the Cosmic Microwave Background (CMB). It is almost homogeneous at present, and remains so as we go back in time. The statement that the Universe is expanding can be taken to mean that any two pieces of the CMB are moving apart. The expansion is homogeneous (the same everywhere) and isotropic (the same in all directions). This means that the distance $r(t)$ between *any* two pieces of the CMB is proportional to a universal quantity $a(t)$, called the scale factor of the Universe. The scale factor, normalized to 1 at the present epoch, is a fundamental property of the homogeneous expanding Universe.

Since $r(t)$ is proportional to $a(t)$ we can define a time-independent quantity x by

$$r(t) = xa(t). \tag{5.1}$$

The quantity x is called the **comoving** distance between the pieces of the CMB. The word comoving is pronounced co-moving and it means 'moving with the expansion'. Quite generally, if any quantity $f(t)$ is proportional to $a(t)$ we write

$$f(t) = fa(t), \tag{5.2}$$

and call f the comoving quantity.

To describe the expansion, I focussed on the CMB. I could as well have focussed on the Cosmic Neutrino Background. I could also have focussed on the baryons between the galaxy clusters, or on the CDM between galaxy clusters. These too constitute an almost homogeneous gas. We have altogether then, a nearly homogeneous gas comprising the CMB, the Cosmic Neutrino Background and (between galaxy clusters) baryons and the CDM.

Each object is moving through the isotropically expanding CMB, with some velocity called its peculiar velocity. In particular, each galaxy has some peculiar velocity. If every galaxy had zero peculiar velocity, the distance between any two of them would be proportional to the scale factor. Taking account of the peculiar velocity, it's still the case that the distance between a pair of galaxies is almost exactly proportional to the scale factor if they are far apart. That is because galaxies far apart are moving away from each other so quickly that their peculiar velocities make hardly any difference. As we will see, the galaxies didn't always exist. Before they formed, the whole Universe was a nearly homogeneous gas, that was expanding almost isotropically.

In addition to the scale factor $a(t)$, it is useful to consider $\dot{a}/a$, which is called the Hubble parameter and denoted by H. It is also useful to consider $1/H$ which is called the Hubble time. If gravity had no effect, $\dot{a}$ would be constant and we would have $a = \dot{a}t$ where t is the age of the Universe starting from the epoch when $a = 0$. In that case, the Hubble time at each epoch would be exactly equal to the age of the Universe. In reality gravity does have an effect, but we shall see in Chapter 8 that the effect is not very big during the Big Bang, so that the Hubble time is roughly equal to the age of the Universe. We shall also see that c/H, called the Hubble distance, is roughly equal to the distance to the horizon defined as the biggest distance that anything could have travelled since the beginning of the Big Bang.

5.3 COLLISION AND DECAY PROCESSES

The known history is summarised in Table 5.1. To understand it, we need to consider the collision processes that occur. The collision processes occurring at the beginning of the known history of the Universe are listed in Table 5.2. In each case, the process can take place in either direction, as is indicated by the double arrows. For each forward process the mass of the initial particles is bigger than that of the final particles, which means that the process releases an amount of kinetic energy that is equal to c^2 times the difference between the masses. The reverse process can occur only if the kinetic energy of the colliding particles is bigger than c^2 times the difference between the initial and final masses.

Later on in the history, other processes become important which are listed in Table 5.3. Again, the initial objects have more mass than the final objects, which means that the mass difference is released as kinetic energy, and that the reverse processes can occur only if the kinetic energy of the incoming particles is bigger than the mass difference. Let us see in more detail what is required. The speed of the nucleus or atom in the early Universe is much less than c. Momentum conservation in this case requires $mv = E/c$ where m and v are the mass and speed of the nucleus or atom and E is the energy of the photon. The kinetic energy of the nucleus or atom is

$$\frac{1}{2}mv^2 \ll mvc = E. \tag{5.3}$$

Therefore practically all of the initial kinetic energy belongs to the photon which means that the energy of the *photon* has to be bigger than c^2 times the mass difference.

The second process in Table 5.3 is an example of nuclear fusion, in which two nuclei combine to form a heavier nucleus with the release of kinetic energy. Nuclear fusion makes the sun and stars hot, and is used in hydrogen bombs. That is in contrast with nuclear fission, which releases kinetic energy by breaking up nuclei. Nuclear fission is used in nuclear reactors and in ordinary atomic bombs.

5.4 THE FIRST MINUTE

Now I'll go through the history stage by stage. As indicated in the first line of Table 5.1, we know that there was thermal equilibrium before neutrino decoupling. We don't know though, for how long before. That is why the *known* history of the Universe begins only when the age is a bit less than 1 s. At this stage the Universe contains several particle species. There are protons, neutrons, photons, electrons, positrons, neutrinos, anti-neutrinos and the CDM. All of the collision processes listed in Table 5.2 are frequently occurring. The processes are in thermal equilibrium, which means that the inverse of each process occurs at the same rate.

The state of thermal equilibrium will be described in Chapter 7. It determines the number of protons per neutron; they are about equal at early times but by the end of the first stage there are about 7 protons for every neutron. Thermal equilibrium also determines the number densities and the energy densities of the other species. It turns out that the number densities of each of the other species are about equal, and so are the energy densities. The number densities and energy densities are much bigger than those of the protons and neutrons.

As the Universe expands, the distance between particles increases and their average energy decreases. In particular, as we see in Chapter 6, the average energy of a photon is proportional to $1/a$. The increasing distance and the decreasing energy means that collision processes involving neutrinos become less likely. When the Universe is about one second old they no longer take place. This is the epoch of neutrino decoupling. The neutrinos then existing, exist still as the Cosmic Neutrino Background.

Soon after neutrino decoupling, the energy of a typical photon becomes less than the rest energy of an electron. As a result, the creation of electron-positron pairs by a pair of photons ceases, and the inverse process of electron-positron annihilation removes the positrons and nearly all the electrons, leaving just one electron per proton.

5.5 BIG BANG NUCLEOSYNTHESIS (BBN)

After about five minutes, the neutrons combine with protons to form nuclei. This is called Big Bang Nucleosynthesis or BBN. It did not happen earlier, because the nuclear collision process in Table 5.3 broke up any nucleus that managed to form.

BBN starts with the formation of ^{2}H nuclei, which is the reverse of the process shown in the first line of Table 5.3. The formation becomes effective only when the average photon energy becomes so small that there are hardly any photons with enough energy to undergo the direct process. More nuclear reactions quickly occur after the formation of ^{2}H nuclei, and Figure 5.1 shows how the calculated abundance of each nucleus (and of the neutron) changes with time. To do this calculation, the rate at which each nuclear reaction occurs has to be determined by laboratory experiments.[2]

Almost all of the neutrons end up inside ^{4}He nuclei, which make up about 25% of the mass of the baryons. Almost all of the remaining 75% comes from single protons (ordinary hydrogen nuclei). Some other light nuclei are created at the same time, and the others are created much later in supernovas, but these account for only a tiny fraction of the mass of the baryons. As a result, about 25% of the mass of the baryons is still ^{4}He and 75% is still ordinary hydrogen.

A tiny fraction of the neutrons end up inside other light nuclei. These are ^{2}H and ^{3}H, ^{3}He, and various lithium and beryllium nuclei. One would like to compare the calculated abundances with observation. What we observe though, are the abundances within stars of various kinds. During the process of star formation, further nuclear reactions take place which affect the abundances. To make the comparison one has to allow for those reactions, which can be done with sufficient accuracy only for ^{4}He, ^{2}H, and ^{7}Li. The reactions hardly affect the abundance of ^{4}He, but they affect the other abundances quite a bit. Allowing for the reactions, the calculated abundances agree with observation, except for the ^{7}Li abundance which seems to be too high.

The lithium discrepancy might be due to an error in the abundance of lithium in stars, that is deduced from observation. It might also be due to an error in the calculation of the change in the lithium abundance, caused by nuclear reactions within those stars. Otherwise, the lithium discrepancy means that there is something wrong with the calculation of the original lithium abundance. There could be an error in the laboratory determinations of the relevant nuclear reaction rates. Instead though, there could be an error in what has been assumed about the early Universe. One possibility is that an unstable particle is present, which decays at about the same time as the lithium is formed. Another is that the density of the protons and neutrons depends strongly on position. A third is that one or more of the fundamental constants is actually varying with time so that it is different when the lithium is formed. Time will tell which explanation is correct so — watch this space!

5.6 LAST SCATTERING AND GALAXY FORMATION

After the formation of light nuclei, nothing happens for about half a million years. Then, almost every electron combines with a nucleus to form a helium or hydrogen atom. This didn't happen earlier, because there were many photons with energy bigger than the atomic binding energies. Those photons would have broken up any atom that managed to form.

After atom formation, the photons no longer bounce off electrons as they did earlier. To understand why, it's useful to think in terms of electromagnetic radiation.

---

[2]The rates can in principle be calculated from the Standard Model, but the calculation is too difficult to perform.

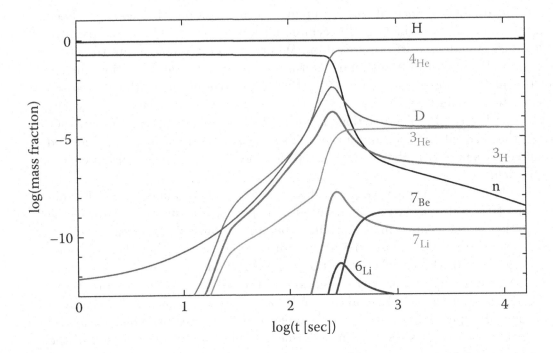

Figure 5.1 Formation of light nuclei The horizontal scale gives the age t of the Universe and the vertical scale gives the mass fraction of each nucleus. The label n denotes the neutron, while H denotes a proton and D denotes a ^{2}H nucleus. When t is less than a few seconds, the number of neutrons per proton is determined by thermal equilibrium. After that, thermal equilibrium fails for the neutrons and protons and the neutron starts to decay with its half-life of 15 minutes. When t is about 300 s, almost all of the neutrons bind into ^{4}He nuclei, but there is a small fraction of other nuclei. (Source: Lawrence Berkeley Laboratory)

After atom formation, each electron is inside an atom and the wavelength of the CMB is much bigger than the size of an atom. Since the atom has zero total charge, the electromagnetic radiation behaves as if it were encountering an object with zero charge which means that nothing happens to it.

Because it marks the end of photon scattering, the epoch of atom formation is usually called **the epoch of last scattering**. The photons existing just after last scattering are now seen as the Cosmic Microwave Background (CMB).

While all of these things are happening, the Universe is nearly homogenous. It's not completely homogeneous though, and some regions are slightly more dense then their surroundings. Additional matter is attracted towards such regions through the force of gravity, and after about 100 million years some of the regions collapse under their own weight, to form the first galaxies. Within each galaxy, over-dense regions in turn collapse to form stars leading to the Universe that we now observe.

age	E	What happens
$< 1\,\mathrm{s}$	$> 1\,\mathrm{MeV}$	Frequent collisions
$1\,\mathrm{s}$	$1\,\mathrm{MeV}$	Neutrino decoupling
		Electron-positron annihilation
$300\,\mathrm{s}$	$0.17\,\mathrm{MeV}$	Big Bang Nucleosynthesis
$380{,}000\,\mathrm{yr}$	$7.0 \times 10^{-7}\,\mathrm{MeV}$	Epoch of last scattering
$1.9 \times 10^{8}\,\mathrm{yr}$	$1.1 \times 10^{-8}\,\mathrm{MeV}$	First galaxies form
$5.8 \times 10^{9}\,\mathrm{yr}$	$1.3 \times 10^{-9}\,\mathrm{MeV}$	Galaxy clusters form
$9.3 \times 10^{9}\,\mathrm{yr}$	$9.2 \times 10^{-10}\,\mathrm{MeV}$	Sun forms
$45.6 \times 10^{9}\,\mathrm{yr}$	$6.5 \times 10^{-10}\,\mathrm{MeV}$	Present time

Table 5.1 **Known history of the Universe** The first column shows the age of the Universe. The second column shows the average energy of a photon, which is proportional to $1/a$ except during electron-positron annihilation. Galaxies and galaxy clusters form over an extended period of time around the one given.

process	k. e. released
$e + \bar{e} \rightleftharpoons \gamma + \gamma$	$1.02\,\mathrm{MeV}$
$e + \bar{e} \rightleftharpoons \nu_\alpha + \bar{\nu}_\alpha$	$1.02\,\mathrm{MeV}$
$n + \nu_e \rightleftharpoons p + e$	$0.79\,\mathrm{MeV}$
$n + \bar{e} \rightleftharpoons p + \bar{\nu}_e$	$1.81\,\mathrm{MeV}$

Table 5.2 **Particle collision processes** In the second line, the neutrino can be any of the three species ν_e, ν_μ or ν_τ.

process	k. e. released
$p + n \rightleftharpoons {}^2H + \gamma$	2.2 MeV
$p + {}^2H \rightleftharpoons {}^3He + \gamma$	5.5 MeV
$p + e \rightleftharpoons \text{hydrogen atom} + \gamma$	13.6 eV

Table 5.3 **Nuclear and atomic collision processes**

EXERCISES

1. It's estimated that there are now roughly 10^{11} galaxies in the observable Universe. Use the information in Table 4.1 and its caption to find the following; (i) the average distance between galaxies, (ii) the average distance between them when they were created at $z \simeq 10$, (iii) the epoch when they would have been in contact had they then existed. Why could they not in fact have existed at this last epoch?

2. Calculate the Hubble distance at the epoch of last scattering. How is that related to the distance to the horizon for an observer at that epoch?

3. Estimate the angle in the sky, subtended by a straight line perpendicular to the line of sight which is tangent to the surface of last scattering and has length equal to the Hubble distance at that epoch.

4. Use the particle masses given in Table A.5 to calculate the amounts of kinetic energy released by the processes in of Table 5.2, confirming the numbers given there.

Energy density of the Universe

CONTENTS

In this chapter I deal with some aspects of the energy density of the Universe. To keep things simple I will ignore the era before electron-positron annihilation, which is dealt with in Chapter 7. Also, I will ignore the neutrino masses because they become important only at a very late stage as described in Section 1.6. With these simplifications the Universe has four components. There are the baryons and the CDM which in the terminology introduced after Eq. (3.3) are matter, and the neutrinos and photons which are radiation.

6.1　THE COSMOLOGICAL REDSHIFT

All galaxies except the nearest few are receding from us. As a result, the frequency of their spectral lines is reduced. The redshift, denoted by z is defined by

$$z = \frac{f - f_{\text{obs}}}{f_{\text{obs}}} \tag{6.1}$$

where f_{obs} is the observed frequency and f is the true frequency, seen be an observer moving with the source.

For $z \ll 1$ the redshift corresponds to a Doppler shift, $z = v/c$ where v is the recession speed of the galaxy (i.e. the rate of increase of its distance).[1] To relate v to the distance of the galaxy, let us pretend that the Earth and the galaxy are both moving with the CMB. Then the distance of the galaxy when the light was emitted

---

[1]For radiation coming from a star in our galaxy or a nearby one, the frequency can be increased or decreased, depending on whether the star is moving towards us or away from us. Doppler thought that this explained why our galaxy contains stars that appear red and others that appear blue. That's not so because their typical speed is only about 0.001 times the speed of light which means that the Doppler shift has no significant effect on the colour. We see each star in its true colour.

was $r(t) = a(t)r_0$ where r_0 is the present distance and t is the time of emission. Its speed was $v(t) = \dot{a}(t)r_0$, therefore $v(t) = H(t)r(t)$.

Since we are dealing with $v \ll c$, we are dealing with $r \ll c/H$. The time t of emission is therefore almost the same as the present time, which means that we have to good accuracy $v_0 = H_0 r_0$ where the subscripts denote the present values. This is called Hubble's law. As shown in Figure 6.1, Hubble's law is not exactly correct, because each galaxy has a peculiar velocity (i.e. it is moving through the CMB).[2]

To make a plot like Figure 6.1, one has to know the Doppler shift of each galaxy *and* its distance. The distance is difficult to determine, and it is known for only a small fraction of the galaxies that have been observed. For the other galaxies, one uses the Doppler shift and Hubble's law to determine the distance, ignoring the effect of random motion.

Using the definition $H = \dot{a}/a$ of the Hubble parameter, and Hubble's law, we can arrive at another interpretation of the redshift:

$$z = \frac{r}{c}\frac{\dot{a}}{a} = t\frac{\dot{a}}{a} \simeq \frac{\delta a}{a}, \tag{6.2}$$

where t is the time taken for the light to arrive and δa is the change in a during that time. This expression becomes exact in the limit $\delta a \to 0$.

Galaxies are also observed with redshift z up to 11. To interpret such big redshift, we should populate the line of sight with an infinite sequence of imaginary observers, who move with the expansion and are infinitesimally close to each other. The redshift going from one observer to the next is a Doppler shift given by Eq. (6.2), and adding them together we find

$$\int_{f_{\text{obs}}}^{f} \frac{df}{f} = \int_{a}^{1} \frac{da}{a}. \tag{6.3}$$

This gives

$$1 + z = \frac{1}{a} \tag{6.4}$$

where a is the scale factor when the radiation was emitted. It is usual to specify an epoch by giving z rather than a.

Although there were no galaxies before $z \simeq 10$ or so, there *was* a gas emitting radiation which will be observed in the future. The best information will come from observation of the 21cm line, corresponding to the transition between the ground state of hydrogen and its first excited state. That will tell us even more than observation of the CMB is telling us now.

Eqs. (6.1) and (6.4) are equivalent to $\lambda/\lambda_{\text{obs}} = a$. In other words, to the statement that an observer at a given epoch sees a wavelength which is proportional to a, i.e. which is expanding with the Universe. With this in mind, it is sometimes said that that the expansion of the Universe means that space is expanding. To put

---

[2]The Earth's peculiar velocity is ignored in making this plot.

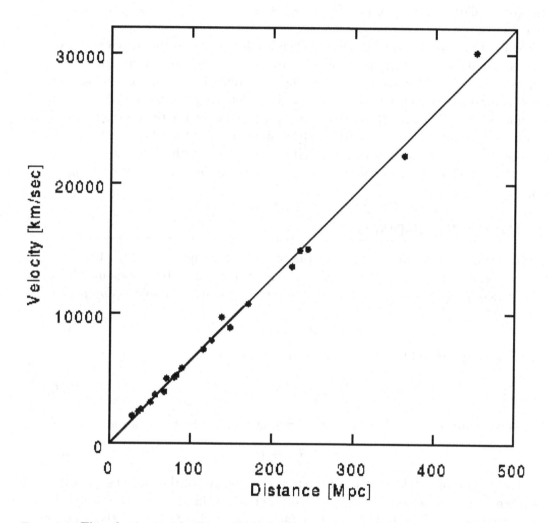

Figure 6.1 This plot is given by Riess, Press & Kirshner in the *Astrophysics Journal* (1996). The dots represent supernovas of a certain type, whose distance can be deduced from their brightness. The straight line corresponds to Hubble's law, assuming that $1/H_0 = 1.53 \times 10^{10}$ yr. Taking into account the limited accuracy of the observation, this is in adequate agreement with the value $1/H_0 = 1.44 \times 10^{10}$ yr, that is obtained from the CMB anisotropy and the galaxy distribution. A plot of this kind was first made by Hubble in 1929, and provided the first evidence for the expansion of the Universe.

it kindly that is misleading, because according to the last two lines of Table 5.1 it would mean that a planet formed soon after the sun, moving in an almost circular orbit, is now nearly twice as far away from the sun as it was initially. According to both Newtonian mechanics and General Relativity, the distance would be the same (or smaller, taking account of energy lost through friction and gravitational radiation).

Finally, I note an important point. Instead of focussing on the wavelength and frequency of an electromagnetic wave, one can focus on the momentum and energy of the corresponding photons. By considering the Lorentz transformation going from one observer in the sequence to the next, one finds that the photon energy is reduced by a fraction v/c. That result applies to any particle moving with speed c and it is a good approximation for any particle whose speed is close to c. We conclude that the energy of any particle moving through the Universe with speed close to c, is proportional to $1/a$ when measured by a sequence of comoving observers along its path.

6.2   MATTER AND RADIATION

In this section I discuss the energy densities of the four components of the Universe. When a region with volume V expands, its energy E changes by an amount $dE = -PdV$ where P is the pressure.[3] The energy density $\rho = E/V$ satisfies $V d\rho + \rho dV = -PdV$, or

$$d\rho = -(\rho + P)dV/V. \tag{6.5}$$

Since V is proportional to a^3 we have $dV/V = 3da/a$ giving

$$a\frac{d\rho}{da} = -3(\rho + P). \tag{6.6}$$

Multiplying both sides by $H = \dot{a}/a$, we arrive at an alternative equation

$$\dot{\rho} = -3H(\rho + P). \tag{6.7}$$

These equations holds for each of the four components of the Universe provided that there is no exchange of energy between them, and that is the case throughout the known history (after electron-positron annihilation). The CDM and the baryons are matter in the terminology introduced after Eq. (3.3), meaning that they have $P \ll \rho$. For matter, Eq. (6.6) gives $\rho \propto a^{-3}$. The photons are radiation meaning that they have $P = \rho/3$. For radiation, Eq. (6.6) gives $\rho \propto 1/a^4$.

As we shall see in Section 7.7, the neutrinos are also radiation to high accuracy if we exclude very late times. With that exclusion we can write the neutrino energy density as

$$\rho_\nu(a) = \rho_{\nu 0}a^{-4}, \tag{6.8}$$

where ρ_{nu0} is what the present neutrino energy density would be if the neutrinos had zero mass.

---

[3]There is no heat flow in a homogeneous Universe.

For each species, it is useful to introduce the energy density fraction

$$\Omega_i \equiv \frac{\rho_{i0}}{\rho_0}, \tag{6.9}$$

where ρ_{i0} is defined by Eq. (6.8) for the neutrinos, and for the other species it is their present energy density. The energy density fraction of each of the four components is shown in Table 6.1, along with the energy density fraction of the vacuum that we come to shortly.

The radiation and matter give equal energy densities at the epoch when

$$1 = \frac{\rho_{\rm b} + \rho_{\rm c}}{\rho_\gamma + \rho_\nu} = \frac{\Omega_{\rm b} + \Omega_{\rm c}}{\Omega_\gamma + \Omega_\nu} a, \tag{6.10}$$

Denoting the redshift at that epoch by $z_{\rm eq}$, Table 6.1 gives $z_{\rm eq} = 3380$. This is well before the epoch $z_{\ell \rm s} = 1100$ of last scattering which means that the Universe is matter dominated at last scattering.

6.3   THE COSMOLOGICAL CONSTANT

Tables 6.1 shows that the biggest contribution to the present energy density, accounting for nearly 70% of the total, comes from the vacuum. The vacuum energy density is called the cosmological constant because it doesn't change with time.

We know that the cosmological constant exists because of its gravitational effect. This is the same as the situation for the CDM and for that reason, the vacuum energy is sometimes called dark energy.

The idea that the vacuum can have energy density is, of course outrageous. Einstein realised almost immediately that this is possible according to General Relativity, but the idea was not popular and a cosmological constant was generally regarded as unlikely.[4] Only in the late 1900s, was the existence of the cosmological constant finally established. Nature has evidently chosen to be outrageous!

Apart from its existence, the cosmological constant is puzzling in another way. As we will see in Section 17.4, a quantum effect, known as a vacuum fluctuation, would by itself make the cosmological constant far bigger than it actually is. To get the observed value of the cosmological constant, we must suppose that it also has a 'classical' contribution, which is negative with magnitude very close to the classical contribution so that two almost cancel.

There is no known reason why the cancellation should occur. On the other hand, if it did not occur, the cosmological constant would be much bigger that it is. As we will see in Chapter 15, that would have prevented the formation of galaxies, which would mean that there would be no stars, and no humans. Partly with that fact in mind, many physicists entertain the idea that the observable Universe may be

---

[4]Einstein himself didn't regard the cosmological constant as the energy density of the vacuum, but instead as a fundamental constant possessed by General Relativity in addition to G and c. That makes no difference, because the effect of Einstein's cosmological constant is exactly the same as the effect of vacuum energy density.

Cosmological Constant Ω_Λ	CDM Ω_c	baryons Ω_b	CMB Ω_γ	Cosmic Neutrino Background Ω_ν
0.68	0.264	0.049	5.4×10^{-5}	3.7×10^{-5}
± 0.01	± 0.001	± 0.001	$\pm 0.1 \times 10^{-5}$	$\pm 0.1 \times 10^{-5}$

Table 6.1 The table shows the fractions of the present energy density, denoted by Ω_i, coming from the different constituents of the Universe. It also shows the fraction coming from the vacuum which is called the cosmological constant. The neutrino masses are set to zero when specifying Ω_ν.

just one of many Universes, in which the cosmological constant takes on practically every value. Then we live in one of the Universes that corresponds to a tiny value simply because we could not live anywhere else.

The same line of reasoning has been invoked to explain some other puzzling features of the world we live in. It is called the anthropic principle and the physics community is polarized between those who love it and those who hate it. Those who hate it take the view that the smallness of the cosmological constant is an accident, or else that it can be explained by some as yet undiscovered theory. That situation is unlikely to change, because the existence of a few coincidences will not convince everybody that the anthropic principle is their explanation.

EXERCISES

1. Show that the 7 protons per neutron present at the epoch of BBN correspond to 25% by weight of ^{4}He and 75% of protons.

2. Calculate the present frequency of a gravitational wave whose wavelength is equal to the Hubble distance at the epoch of matter-radiation equality.

3. How many neutrino species would be needed, to make the epoch of matter-radiation equality coincide with the epoch of last scattering?

4. Estimate the slope of the line in Figure 6.1. Use your estimate to verify that the slope corresponds to $1/H_0 \simeq 1.5 \times 10^{10}$ yr.

Thermal equilibrium

CONTENTS

A process in the early Universe is in thermal equilibrium, only if the rate per particle is much bigger than the Hubble parameter. This is because a typical particle in the opposite case would not interact within a Hubble time, which because of the expansion of the Universe means that it would not interact at all. In this chapter I describe thermal equilibrium in general, and then see how it applies in the early Universe. I will use natural units, and set the Boltzmann constant to 1 so that T is measured in eV.

The discussion will not apply to the CDM because it rarely if ever undergoes collision processes with baryonic matter and will not be in thermal equilibrium with it. The CDM particles might be bouncing off each other, which could put them into thermal equilibrium at a much lower temperature. Such 'self-interacting CDM' is not thought to be likely though, which means that the CDM is probably not in thermal equilibrium even with itself.

7.1   DISTRIBUTION FUNCTIONS

The properties of a gas are determined by the occupation numbers of the quantum states corresponding to definite momentum and spin for each constituent.[1] I will denote these by f_s where s runs over the labels of the constituents.

Consider now a process $a + b \rightarrow c + d$ where the letters denote particle species. For simplicity I will take the two final particles to be different, and because I am going to consider the inverse process I will take the two initial particles to be different as well. Quantum physics describes the complication that ensues if the two

---

[1] The occupation number is the average number of the constituents per quantum state.

final particles in a process are identical, and it makes no difference to the outcome of the following discussion.

Let us demand that the momentum of each particle is within some element d^3p of momentum space, and ask how the rate depends on the occupation numbers of the particles. From very general principles, it is equal to some constant times $f_a f_b (1 \pm f_c)(1 \pm f_d)$, with plus for a boson and minus for a fermion. The first two factors just account for the abundance of the incoming particles. The minus sign accounts for the fact that no two fermions can occupy the same quantum state. The plus sign is the enhancement that corresponds to stimulated emission in a decay process.

Assuming what is called time reversal invariance, the rate for the reverse process is equal to the same constant times $f_c f_d (1 \pm f_a)(1 \pm f_b)$. Time reversal invariance means that collision processes do not specify the arrow of time; if we replace t by $-t$ the description of these processes remains the same. Tiny departures from time reversal invariance are observed, but they are far too small to affect the following discussion.[2]

In thermal equilibrium the rates are the same which means that the occupation numbers satisfy

$$(f_a^{-1} \pm 1)(f_b^{-1} \pm 1) = (f_c^{-1} \pm 1)(f_d^{-1} \pm 1) \qquad (7.1)$$

If we write for each constituent

$$f_s^{-1} \pm 1 = \exp(X_s), \qquad (7.2)$$

Eq. (7.1) becomes

$$X_a + X_b = X_c + X_d. \qquad (7.3)$$

Suppose now that there is thermal equilibrium at temperature T, with $T = 0$ corresponding to absolute zero. Then

$$X_s(E) = \frac{E - \mu_s}{T}, \qquad (7.4)$$

where E is the energy of the particle and μ_s is its chemical potential.[3] As with energy, the total chemical potential is conserved in each process. Reverting to the occupation number Eq. (7.3) becomes (with now a minus sign for a boson and a plus sign for a fermion)

$$f_s(E) = \frac{1}{\exp\left(\frac{E - \mu_s}{T}\right) \pm 1}. \qquad (7.5)$$

The occupation number of the constituent determines the distribution $n_s(p)$ of

---

[2] To be precise, one observes a tiny departure from invariance under what is called the PC transformation. That implies a departure from invariance under time reversal T because according to the PCT theorem every field theory is invariant under the combined transformations PC and T.

[3] The minus in front of μ_b is a convention. Given that, the appearance of $E - \mu_s$ up to an overall constant (with μ_s conserved) is demanded by the conservation of X_s. Identifying the constant with $1/T$ may be regarded as a definition of temperature.

number density and the distribution $\rho_s(p)$ of energy density, such that $n_s(p)d^3p$ and $\rho_s(p)d^3p$ are the number- and energy densities of the constituent that have momentum p within a cell d^3p. Remembering that $(2\pi)^{-3}d^3p$ is the density of quantum states with given spin and momentum in a cell d^3p, these are given by

$$n_s(p) = \frac{g_s}{(2\pi)^3} f_s(E) \tag{7.6}$$

$$\rho_s(p) = \frac{g_s}{(2\pi)^3} f_s(E)E, \tag{7.7}$$

where g_s is the number of spin states and $E(p) = \sqrt{p^2 + m^2}$ is the energy of a constituent with mass m.

7.2 GENERALISED BLACKBODY DISTRIBUTIONS

If the mass and chemical potential of a constituent are much less than T, it is a good approximation to set both of them to zero. Then Eqs. (7.6) and (7.7) become

$$n_s(p) = \frac{g_s}{(2\pi)^3} \frac{E^2}{\exp\left(\frac{E}{T}\right) \pm 1} \tag{7.8}$$

$$\rho_s(p) = \frac{g_s}{(2\pi)^3} \frac{E^3}{\exp\left(\frac{E}{T}\right) \pm 1}, \tag{7.9}$$

with $E = p$. Photons in thermal equilibrium with charged particles satisfy Eqs. (7.8) and (7.9) (with $g_s = 2$ for the two photon helicities). Indeed, photons have zero mass, and charged particles can produce photons when they collide without changing their identity which means that the photons also have zero chemical potential. As Eqs. (7.8) and (7.9) for photons correspond to the blackbody distribution, one may call these equations the generalised blackbody distribution.

For the generalised blackbody distribution, the number- and energy densities are given by

$$n_s = \frac{g_s}{2\pi^2} \int_0^\infty f_a(E)E^2 dE = \frac{2\zeta(3)g_s}{\pi^2} T^3 \times \left\{ \begin{array}{ll} 1 & \text{(bosons)} \\ 3/4 & \text{(fermions)} \end{array} \right., \tag{7.10}$$

where $\zeta(3) = 1.202\ldots$, and

$$\rho_s = \frac{g_s}{2\pi^2} \int_0^\infty f_a(E)E^3 dp = \frac{2\pi^2 g_s}{30} T^4 \times \left\{ \begin{array}{ll} 1 & \text{(bosons)} \\ 7/8 & \text{(fermions)} \end{array} \right.. \tag{7.11}$$

We see that constituents with the generalised blackbody distribution all have roughly the same number- and energy densities. Their average particle energy ρ_s/n_s is roughly equal to T which means that it is much bigger than the particle mass. In the terminology of Chapter 5, these constituents are therefore radiation as opposed to matter.

7.3 MAXWELL-BOLTZMANN DISTRIBUTION

If T is much less than m_s, we deal with a matter species. Then, to a good approximation Eq. (7.5) becomes

$$f_s = \exp\left(\frac{\mu_s - p^2/2m_s}{T}\right),$$ (7.12)

which corresponds to

$$n_s(p) = \frac{g_s}{(2\pi)^3}\exp\left(\frac{\mu_s - p^2/2m_s}{T}\right).$$ (7.13)

This gives

$$n_s = \frac{g_s}{2\pi^2}\exp\left(\frac{\mu_s - m_s}{T}\right)(m_sT)^{3/2}\int_0^\infty \exp\left(-\frac{1}{2}x^2\right)x^2 dx$$ (7.14)

$$= g_s\exp\left(\frac{\mu_a - m_a}{T}\right)\left(\frac{m_aT}{2\pi}\right)^{3/2}.$$ (7.15)

The fraction $n_s(p)/n_s$ is the Maxwell-Boltzmann distribution.

7.4 INITIAL THERMAL EQUILIBRIUM

Now we deal with the first stage of the known history, when the constituents are particles and all except the CDM are in thermal equilibrium at a temperature T which is a bit bigger than 1 MeV. We first need to specify the number of baryons per photon, which is constant during the first stage because baryon number conservation means that the density of baryon number is proportional to $1/a^3$ just like the number density of the photons. As we will see shortly, the baryon number per photon during the first stage is 4/11 times its present value. From observation, we learn that the latter is 6.1×10^{-10}, which means that the baryon number per photon during the first stage is 2.2×10^{-10}. The smallness of the baryon number per photon is very important for cosmology.

Next, we need to consider in turn the collision process listed in Table 5.2. The process in the first line requires that the electron and positron chemical potentials satisfy $\mu_e = -\mu_{\bar{e}}$. To make the Universe neutral we need an electron excess of one per proton. That requires $\mu_{\bar{e}}$ to be positive and much less than T, so that Eqs. (7.10) and (7.11) apply to the electrons and positrons.

The second process in Table 5.2 requires $\mu_\nu = -\mu_{\bar{\nu}}$ for each neutrino species. One usually assumes that $|\mu_\nu| \lesssim \mu_{\bar{e}}$ for each neutrino species, which corresponds to the statement that the density of each lepton number is at most on the order of the density of baryon number. In any case, successful Big Bang Nucleosynthesis requires $|\mu_\nu| \ll T$ so that again Eqs. (7.10) and (7.11) apply with high accuracy.

Adding the contributions, we find the total energy density of the radiation;

$$\rho_r = \frac{g_*\pi^2}{30}T^4,$$ (7.16)

where[4]

$$g_* = 2 + \left(5 \times 2 \times \frac{7}{8}\right) = 10.75. \qquad (7.17)$$

Since the baryon number per photon is very small, the radiation cannot exchange a significant amount of energy with the baryons. As we saw in Chapter 5, this means that its energy density is proportional to $1/a^4$. Since it is also proportional to T^4, it follows that T is proportional to $1/a$. This has an important implication for the distributions of number- and energy density, $n_s(E)dE$ and $\rho_s(E)dE$ given by Eqs. (7.8) and (7.9). Focussing on those particles with energy in an interval E to $E + dE$, with $E \propto T$ and $dE \propto T$, we see that their number density is proportional to $1/a^3$ just as if they were not interacting. We also see that their energy is proportional to $1/a$, which as seen in Section 6.1 is again the same as if they were not interacting. It therefore follows that the distributions (7.8) and (7.9) for the neutrinos remains valid even after they cease to interact, with $T \propto 1/a$. Much later, when the photons cease to interact at the epoch of last scattering, it means that the blackbody distribution also remain valid with again $T \propto 1/a$.

Coming to the processes in the third and fourth lines of Table 5.2, the fact that $\mu_{\bar{e}}$ and $|\mu_\nu|$ are much less than T means that $|\mu_n - \mu_p|$ is also much less than T. As T is much less than m_p, we have $E_p = m_p$ and $E_n = m_n$ to high accuracy which gives

$$\frac{n_n}{n_p} = e^{-(m_n - m_p)/T}. \qquad (7.18)$$

At early times this ratio is close to 1, but at the end of the first stage when thermal equilibrium fails it has fallen to about $1/7$.

In Table A.3 I give five fundamental cosmological parameters, that specify the known history of the homogeneous Universe. We can see why the number is five, by considering the situation during the first stage of the history. To specify the state of the particles in thermal equilibrium during the first stage, we need to specify the temperature at some chosen epoch, and the baryon number per photon. We also need to specify at that epoch the Hubble parameter, the energy density of the CDM, and the cosmological constant, making five in all. These five parameters determine those in Table A.3 and vice versa, and so it makes no difference which set is chosen.

7.5 ELECTRON-POSITRON ANNIHILATION

The photon temperature is the same as the neutrino temperature, until the epoch of electron-positron annihilation which marks the end of the first stage. At this epoch $T \simeq 0.5\,\text{MeV}$ corresponding to the fact that the energy of a typical photon is about equal to the rest energy of the electron. The annihilation has no effect on the neutrino temperature T_ν, which continues to be proportional to a. To calculate

---

[4]The first term is the photon contribution, and the five contributions to the second term are the contributions of the electron, the positron, and the three neutrino species counting both the neutrino and anti-neutrino. All of these contributions have two spin states.

its effect on the photon temperature we need the fact that entropy is conserved during thermal equilibrium. The entropy density s of the photons, electrons and positrons is therefore proportional to $1/a^3$ throughout the era of electron-positron annihilation, which is in turn proportional to $1/T_\nu^3$.

To calculate s, we need the relation $dE = TdS - PdV$, which holds for arbitrary changes in the volume V, the entropy S and the energy E of gas constituents in thermal equilibrium. Writing $E = \tilde{\rho}V$ for the energy of the particles in thermal equilibrium, and using the entropy density $s = S/V$, this becomes

$$d\tilde{\rho} = \left(sT - \tilde{\rho} - \tilde{P} \right) \frac{dV}{V} + Tds, \tag{7.19}$$

where $\tilde{P} = \tilde{\rho}/3$ is the pressure. But s, $\tilde{\rho}$ and $\tilde{P}$ depend only on the photon temperature T because the occupation numbers do, which means that there is in fact no dependence on the volume. The entropy density of the photons, electrons and positrons is therefore $s = (\tilde{\rho} + \tilde{P})/T$. Adding the contributions to $\tilde{\rho}$, it is given by Eq. (7.16) with $g_* = 11/2$ before electron-positron annihilation and $g_* = 2$ after electron-positron annihilation. Since s is proportional to $1/T_\nu^3$, we conclude that after electron-positron annihilation, $T = (11/4)^{1/3} T_\nu$. The energy density is now given by Eq. (7.16) with g_* replaced by an effective g_*^{eff} given by

$$g_*^{\text{eff}} = 2 + \left[\frac{7}{8} \times 6 \times \left(\frac{4}{11} \right)^{4/3} \right] = 3.36. \tag{7.20}$$

7.6 EPOCH OF LAST SCATTERING

To calculate the temperature at last scattering, we need the blackbody distribution $n_\gamma(E)$, given by Eq. (7.8) with the minus sign. For a rough estimate of the temperature at last scattering, we can use the fact that a photon capable of breaking up a hydrogen atom (ionizing photon) must have energy bigger than the binding energy of hydrogen which is $13\,\text{eV}$. Last scattering will take place at roughly the epoch when the number density of ionizing photons falls below the number density of baryons.[5] Because there are so many photons per baryon, that does not happen until T (roughly equal to the *average* photon energy) is well below $13\,\text{eV}$. Let us see how far below it must be. In the regime $E \gg T$, the number density of photons

---

[5] Only the protons are actually relevant, but most of the baryons are protons.

with energy bigger than E is

$$n_\gamma(> E) = \frac{2}{2\pi^2} \int_E^\infty dE \frac{E^2}{e^{\frac{E}{T}} - 1} \qquad (7.21)$$

$$\simeq \frac{2}{2\pi^2} \int_E^\infty dE E^2 e^{-\frac{E}{T}} \qquad (7.22)$$

$$\simeq \frac{2T^3}{2\pi^2} \int_{\frac{E}{T}}^\infty dx x^2 e^{-x} \qquad (7.23)$$

$$\simeq \frac{2T^3}{2\pi^2} \left(\frac{E}{T}\right)^2 e^{-\frac{E}{T}}. \qquad (7.24)$$

This gives

$$\frac{n_\gamma(> E)}{n_B} = \frac{1}{6 \times 10^{-10}} \frac{n_\gamma(> E)}{n_\gamma} \simeq \frac{1}{6 \times 10^{-10}} \left(\frac{E}{T}\right)^2 e^{-\frac{E}{T}}, \qquad (7.25)$$

which is equal to 1 when $T = E/27$. Setting $E = 13\,\text{eV}$, we learn that there is one ionizing photon per baryon when $T = (13\,\text{eV})/27.3 = 0.47\,\text{eV}$, which is our estimate for the photon temperature at last scattering. The present temperature is $T_0 = 2.75\,°\text{K}$ which in our units becomes $T_0 = 2.36 \times 10^{-4}\,\text{eV}$, According to this estimate, last scattering therefore takes place at $z \simeq 0.47/T_0 = 1980$.

For a more accurate result, we need to calculate the fraction of free protons remaining at a given temperature;

$$X \equiv \frac{n_p}{n_p + n_H} \qquad (7.26)$$

$$\simeq \frac{n_p}{n_b} \qquad (7.27)$$

where p indicates a free proton and H a hydrogen atom and the second expression ignores the one neutron per proton. The epoch of last scattering is by convention defined as the one when X, taken to be given by the second expression, has fallen to 0.1.

To calculate X we need the equilibrium number densities given by Eq. (7.15). Since the processes $p + e \leftrightarrow H$ are occurring, $\mu_p + \mu_e = \mu_H$. We have therefore (setting $m_p/m_H = 1$)

$$\frac{n_H}{n_e n_p} = \frac{g_H}{g_e g_p} \left(\frac{2\pi}{m_e T}\right)^{3/2} \exp\left(\frac{E_b}{T}\right), \qquad (7.28)$$

where $E_b = 13\,\text{eV}$ is the hydrogen binding energy. We have $g_e = g_p = 2$ (two spin states each). The H atom is supposed to be in the ground state (orbital angular momentum zero) so it has $g_H = 4$ corresponding to the four spin states. Also, electrical neutrality requires $n_e = n_p$.

Using Eq. (7.28), together with Eq. (7.10) for n_γ, we find a useful expression for X as given by Eq. (7.27);

$$\frac{1 - X}{X^2} = \frac{n_b}{n_\gamma} \frac{n_H n_\gamma}{n_p^2} \tag{7.29}$$

$$= 3.82 \frac{n_B}{n_\gamma} \left(\frac{T}{m_e} \right)^{3/2} e^{E_b/T}. \tag{7.30}$$

Taking last scattering to be $X = 0.1$, this gives at last scattering $T = 0.31\,\text{eV}$. Reverting to degrees Kelvin this corresponds to $T = 3600\,°\text{K}$ and therefore to redshift

$$z_{ls} = 3600/2.7 \simeq 1100. \tag{7.31}$$

7.7   COSMIC NEUTRINO BACKGROUND

The Cosmic Neutrino Background was created at the epoch of neutrino decoupling. At that stage the energy of each neutrino was much bigger than its rest energy. Repeating the discussion that was just given for the energy of a photon, we see that the energy of a neutrino decreased like $1/a$, unless and until it becomes comparable with the rest energy. Let us see whether that happens for one or more of the neutrino species. From Eqs. (7.10) and (7.11) one sees that the average energy of a neutrino is about the same as the average energy of a photon, and that the present average energy of a photon is $3.15 T_0 = 7.43 \times 10^{-4}\,\text{eV}$. Using the rest energies in Table A.5, one sees that the mass of the lightest neutrino might still be negligible at present, but that the masses of the other two neutrinos are not.

Calculating the present neutrino number density from Eq. (7.10) with $T^3 = T_\nu^3 = (4/11) T_0^3$, and invoking the present energy density $\rho_0 = 4840\,\text{MeV}\,\text{m}^{-3}$, one finds that neutrino matter accounts for a fraction $M/46\,\text{eV}$ of the present energy density, where M is the sum of the neutrino masses. If the neutrino masses have the minimum values allowed by observation, this is only 0.0013, which is much less than the fraction 0.048 coming from baryonic matter (see Table 6.1). In that case, neutrino masses have practically no cosmological effect.

If the neutrino masses are bigger, they might have a cosmological effect. By requiring that the effect is not so big that it disagrees with observation, one finds that M must be less than about $1\,\text{eV}$. Equivalently, one finds that the neutrinos can account for at most 0.02 of the present energy density.

EXERCISES

1. Show that Eq. (7.1) follows from the preceding discussion.

2. Show that $n_s(E)$ and $\rho_s(E)$ given by Eqs. (7.8) and (7.9) both peak at $E \sim T$ with the peak width on the order of T. Show that as a result, we need $m_s \ll T$ for the mass of the species, if Eqs. (7.8) and (7.9) are to be good approximations.

3. The CMB has the blackbody spectrum (7.9) with $T = 2.725\,\mathrm{K}$. Use Eq. (7.10) and Table A.3 to show that there are 4.11×10^4 CMB photons per m^3.

4. Use Eq. (7.10) and the result of the previous exercise, to show that there are 3.36×10^4 Cosmic Neutrino Background neutrinos per m^3. Then show that neutrino matter accounts for a fraction $M/46\,\mathrm{eV}$ of the present energy density $\rho_0 = 4{,}850\,\mathrm{m}^{-3}$, where M is the sum of the neutrino masses.

5. Use Eqs. (7.10) and (7.11) to calculate the average energy per photon of photons in thermal equilibrium, in units of T.

6. Using the observed value of n_B/n_γ, verify the statement made after Eq. (7.30).

Friedmann equation

CONTENTS

Chapter 6 described the energy density of the Universe, and its dependence on the scale factor. Chapter 7 described thermal equilibrium, and the dependence of the temperature on the scale factor. Now we need to see how the scale factor depends on the age of the Universe. For that purpose, we need the Friedmann equation, which was derived by Alexander Friedmann in 1922. I continue to set $c = 1$.

8.1   FRIEDMANN EQUATION

The Friedmann equation corresponds to the time-time component of the Einstein field equation (3.9), with the pretense that the Universe is perfectly homogeneous and isotropic. To derive the Friedmann equation in its simplest form, we can write the line element for the homogeneous isotropic Universe in the form

$$ds^2 = -dt^2 + a^2(t)d\ell^2 \tag{8.1}$$
$$d\ell^2 = \delta_{ij}dx^i dx^j \tag{8.2}$$
$$= dx^2 + x^2\left(d\theta^2 + \sin^2\theta d\phi\right). \tag{8.3}$$

I have written the spatial line element first in Cartesian coordinates, and then in spherical polar coordinates.

The worldlines with fixed x^i are moving with the expansion of the Universe, and the Universe is homogeneous in spacetime directions orthogonal to the worldlines corresponding to constant x^i. The worldlines are free-falling because there is no pressure gradient. As a result, nearby worldlines can belong to a locally inertial frame and their clocks can be permanently synchronised, reading the time t which one may call the cosmic time.

After using Eqs. (8.1) and (8.2) to calculate the curvature tensor, the time-time component of the Einstein field equation gives us the simplest form of the Friedmann equation;

$$H^2 = \frac{8\pi G\rho}{3}. \tag{8.4}$$

This is not the most general form of the Friedmann equation because it assumes the existence of Cartesian coordinates. That is equivalent to assuming that the geometry of the Universe is the one corresponding to everyday experience, which was formalised by Euclid around 300 BC. As we see in the next chapter, Euclid's geometry might not apply to the Universe. To be precise, we might have, instead of Eq. (8.3)

$$dl^2 = \frac{dx^2}{1 - Kx^2} + x^2 \left(d\theta^2 + \sin^2 \theta d\phi \right), \tag{8.5}$$

with the constant K either positive or negative. After evaluating the curvature tensor for this metric, the time-time component of the field equation gives the most general form of the Friedmann equation;

$$H^2 = \frac{8\pi G\rho}{3} - \frac{K}{a^2}. \tag{8.6}$$

According to observation, K/H^2 is less than 10^{-2} at the present epoch. Since the present size of the observable Universe is roughly H^{-2}, this implies Kx^2 is much less than 1 throughout the observable Universe. In other words, it implies that the geometry of the observable Universe is almost Euclidean. I will therefore use the simplest form of the Friedmann equation, Eq. (8.4) corresponding to $K = 0$. This gives the energy density in terms of H. At present, $1/H_0 = 14.5\,\mathrm{Gyr}$ which corresponds to $4.76\,\mathrm{GeV\,m}^{-3}$.

Whether or not K vanishes, Eq. (8.6) provides a useful expression for the acceleration of the scale factor:

$$\ddot{a} = \frac{4\pi G}{3} \frac{d}{da} \left(a^2 \rho(a) \right). \tag{8.7}$$

Matter and radiation give a negative contribution to $\ddot{a}$, but the cosmological constant gives a positive contribution,

$$\ddot{a}_\Lambda = \frac{8\pi G}{3} \rho_\Lambda a. \tag{8.8}$$

From Eq. (8.7) follows another useful equation,

$$\dot{H} + H^2 = \frac{4\pi G}{3} \frac{1}{a} \frac{d}{da} \left(a^2 \rho(a) \right). \tag{8.9}$$

8.2 EVOLUTION OF THE SCALE FACTOR

The Friedmann equation (8.4) gives

$$\frac{da}{dt} = \sqrt{\frac{3}{8\pi G}} a \rho^{1/2}(a). \tag{8.10}$$

The time since the beginning of the Big Bang is therefore

$$t(a) = \sqrt{\frac{3}{8\pi G}} \int_{a_{\mathrm{bb}}}^{a} \left[a\rho^{1/2}(a) \right]^{-1}, \tag{8.11}$$

where a_{bb} is the scale factor at the beginning of the Big Bang. Except during the first Hubble time or so, it is a good approximation to set $a_{bb} = 0$. Then, during radiation domination we have $\rho \propto 1/a^4$ giving $a \propto t^{1/2}$ and $H = 1/2t$. During matter domination before the cosmological constant becomes significant we have $\rho \propto 1/a^3$ giving $a \propto t^{2/3}$ and $H = 2/3t$. (These results apply to any era of radiation- or matter domination, excluding an initial era when the time since the beginning is much less than the value of $1/H$ at the beginning.)

After the cosmological constant becomes significant we have

$$\rho = \rho_0 \left(\Omega(a) + (1 - \Omega_0) \right), \tag{8.12}$$

where 0 denotes the present and $\Omega(a)$ is the matter energy density as a fraction of the total, given by

$$\Omega(a) = \Omega_0/a^3, \qquad \Omega_0 = 0.31. \tag{8.13}$$

This gives

$$t(a) = \frac{2}{3H} \frac{1}{\sqrt{1 - \Omega}} \ln \left(\frac{1 + \sqrt{1 - \Omega}}{\sqrt{\Omega}} \right). \tag{8.14}$$

The corresponding $a(t)$ is shown in Figure 8.1. The present age is $t_0 = 1.38 \times 10^{10}$ yr which is close to the Hubble time $1/H_0 = 1.44 \times 10^{10}$ yr.

The Friedmann equation (8.4) also gives us the time dependence of T during radiation domination. During the first stage of the history, ρ is given by Eq. (7.16). Putting this into Eq. (8.4) and remembering that $H = 1/t$ gives

$$\frac{1}{4t^2} = \frac{\pi^2}{90} g_* \frac{(kT)^4}{8\pi G}. \tag{8.15}$$

This is in natural units. Using Table A.3 we can convert G to MKS and t to seconds, which gives

$$\frac{t}{1\,\mathrm{s}} = 2.43 g_*^{-1/2} \left(\frac{1\,\mathrm{MeV}}{kT} \right)^2 ., \tag{8.16}$$

with $g_* = 10.75$. After electron-positron annihilation, this expression holds with g_* replaced by $g_*^{\mathrm{eff}} = 3.36$ as in Eq. (7.20).

With the timescale in place, we can estimate the temperature at neutrino decoupling. Neutrino decoupling occurs when the rate per particle for the neutrino processes in Table 5.2 falls below the Hubble parameter.[1] Using natural units the cross sections σ for these processes are of order $G_F^2 E^2$, where

$$G_F = \frac{\sqrt{2}}{8} \frac{g^2}{m_W^2} = 1.16 \times 10^{-5}\,\mathrm{GeV}^{-2}. \tag{8.17}$$

The collision rate for a given incoming particle moving with speed $c = 1$ is $\Gamma = \sigma n$

---

[1] Very soon after that epoch, a typical neutrino travels freely until the present epoch.

where n is the number density of the other incoming particle. Since $n \sim E^3$ we have $\Gamma \sim G_F E^5$. Using $3M_P^2 H^2 = \rho$, this gives the ratio

$$\frac{\Gamma}{H} \sim M_P G_F^2 (k_B T)^3. \tag{8.18}$$

Setting this equal to 1 gives an estimate for the temperature at the epoch of neutrino decoupling,

$$(k_B T)_{\text{decoup}} \sim (M_P G_F^2)^{-1/3} \sim 1 \, \text{MeV}. \tag{8.19}$$

Using Eq. (8.16) this corresponds to $t_{\text{decoup}} \sim 1 \, \text{s}$.

8.3 DISTANCE TO THE HORIZON

At any epoch, the distance to the horizon is defined as the maximum distance that any particle could have travelled since the beginning of the Big Bang. In other words, as the distance that a particle moving in a straight line with speed c would have travelled. Since the time since the beginning of the Big Bang is about equal to the Hubble time $1/H$, we may expect that the distance to the horizon is about c/H, ie. $1/H$ with our choice $c = 1$. That expectation though, ignores the expansion of the Universe. To verify it, we have to take the expansion into account.

To do that, we need comoving coordinates $\mathbf{x} \equiv \mathbf{r}/a(t)$, instead of the physical coordinates $\mathbf{r}$. Suppose that a photon arrives at the origin $\mathbf{x} = 0$ at some time t, and was at a very short distance $a(t)dx$ a very short time dt earlier. Then $a(t)dx = cdt$ is a good approximation because a hardly changes during the time dt. As the choice of origin is arbitrary, this relation is valid without qualification. Going to infinitesimals we get an exact result, and integrating gives for the comoving distance to the horizon[2]

$$x(t) = \int_{t_{bb}}^{t} dt'/a(t') = \int_{t_{bb}}^{a(t)} da/a^2 H(a) \tag{8.20}$$

$$= \frac{1}{H(t)} \int_{t_{bb}}^{a(t)} \left[\frac{\rho(a(t))}{\rho(a)} \right]^{1/2} \frac{da}{a^2} \tag{8.21}$$

where bb denotes the start of the Big Bang and I have used $\rho \propto H^2$.

Let us first suppose that the Big Bang starts only at the beginning of the known history. During radiation domination when ρ is proportional to $1/a^4$, we can to a good approximation set a_{bb} to zero (except during the first few Hubble times) which gives $r(t) = 1/H(t)$. During matter domination, we can ignore the radiation dominated era (except during the first few Hubble times) which gives to a good approximation $r(t) = 2/H(t)$. At the present epoch we need to keep both the cosmological constant and the matter, but we can still drop the radiation and set a_{bb} to zero. This gives $r_0 = 3.17/H_0$.

---

[2]As in the previous discussion I exclude the initial part of each era, when the time since the beginning is much less than $1/H$ at the beginning.

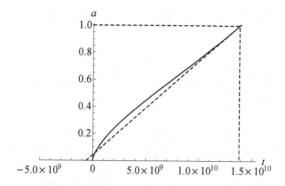

Figure 8.1 The scale factor with t in years. The sloped dashed line is what the scale factor would be, if $\dot{a}$ had its present value at all times. The age of the Universe would then be equal to the present Hubble time $1/H_0$ which is 1.44×10^{10} yr. This is almost the same as actual age $t_0 = 1.38 \times 10^{10}$ yr.

These results are practically unchanged if the Big Bang started earlier, because during the Big Bang the Universe contains matter, radiation or a mixture of the two. Therefore, the integral converges rapidly as we make the start of the Big Bang earlier, giving indeed a negligible change to the result. Also, the result $r = 1/H$ holds during any era of radiation domination and the result $r = 2/H$ holds during any era of matter domination after a few Hubble times have elapsed. We see that in all cases, the distance to the horizon is at least roughly equal to the Hubble distance.

EXERCISES

1. For the metric defined by Eqs. (8.1) and (8.2), show that the nonzero components of the Christoffel symbol are $\Gamma^0_{ij} = \delta_{ij}\dot{a}a$ and $\Gamma^i_{0j} = \Gamma^i_{j0} = H$. Then show that $R_{00} = -3\ddot{a}/a$ and $R = 6(\ddot{a}/a + H^2)$, and put these into the Einstein field equation to verify the Friedmann equation.

2. Calculate the Hubble distance when $T = 1\,\text{MeV}$, using Appendix B.

3. Use the results in Section 8.2 to verify the relationships between the age of the Universe and the scale factor shown in Table 5.1.

4. Verify Eqs. (8.7) and (8.9).

5. Verify Eq. (8.14) by differentiating it with respect to a, and substituting the result into Eq. (8).

6. Einstein pointed out that according to Eq. (8.7) the Universe could be static (constant sca factor) for a suitably chosen cosmological constant ρ_Λ. Calculate the value of ρ_Λ required to make the present Universe static. (These calculations take the Universe to be perfectly homogeneous. With the inhomogeneity

taken into account the static Universe becomes unstable, as was pointed out by Eddington.)

7. Generalise Eq. (8.11) so that it gives the time since an object with redshift z emitted its light. Use your result to calculate the time since an object at $z = 9$ emitted its light, ignoring the cosmological constant and the radiation.

8. Use Eq. (8.21) to calculate the present distance of an object with $z = 9$. Divide this distance by c to give the time since the object emitted its light that would apply if the expansion of the Universe were ignored. Compare it with the time that you calculated in the preceding exercise.

9. Calculate the age of the Universe that would be implied by the observed Hubble time, if the cosmological constant vanished.

10. Differentiate Eq. (8.14) to verify that it satisfies Eq. (8.10). What determines the constant of integration?

The geometry of the Universe

CONTENTS

The central assumption of Euclid's geometry, is the existence of parallel straight lines which indeed accords with everyday experience. In the eighteenth century though, it was shown that this assumption could be altered, to produce two different but still logically consistent geometries. Following Riemann one can assume that every pair of straight lines cross.[1] Alternatively, following Bolyai and Lobachevsky, one can assume that when extended far enough, straight lines will always move apart.

In this chapter, I will first show that any homogeneous geometry must correspond to a line element of the form (8.5) for *some* K. Then I will show that *all* K give a homogeneous geometry. Positive K corresponds to Riemann's geometry while negative K corresponds to that of Bolyai and Lobachevsky. Since K is zero within the observational errors, the chapter may be of only academic interest and will not be used in the rest of the book.

9.1   SPATIAL METRIC

To specify the geometry, we can give an expression for the distance squared $d\ell^2$ between two nearby points in space. Euclidean geometry corresponds to the existence of Cartesian coordinates x^i, such that

$$d\ell^2 = \delta_{ij}dx^i dx^j. \tag{9.1}$$

The most general geometry would correspond to a spatial metric g_{ij} defined by

$$d\ell^2 = g_{ij}dx^i dx^j. \tag{9.2}$$

---

[1] That is analogous to the situation for geodesics on the surface of the Earth. They always cross, and the sum of the angles of a geodesic triangle is bigger than 180°.

Transforming to new coordinates x'^i we get g'_{ij} given by

$$g_{ij} = \frac{\partial x'^n}{\partial x^i}\frac{\partial x'^m}{\partial x^j}g'_{nm}. \tag{9.3}$$

For the geometry to make sense, we need to assume that at a given point one can choose $g_{ij} = \delta_{ij}$. Analogously with the local flatness theorem of Chapter 2, one can show that it is possible to also have $\partial g_{ij}/\partial dx^n$ at the point. This explains why Euclidean geometry is the one that corresponds to everyday experience. It is because everyday experience deals with only small regions of space, within which it is possible to find coordinates such that $g_{ij} = \delta_{ij}$ to high accuracy.

9.2   HOMOGENEOUS AND ISOTROPIC METRIC

In the context of cosmology, we are interested in a spatial geometry which is homogeneous and isotropic. What exactly does that mean? It means that if we use coordinates that have an origin $x^1 = x^2 = x^3 = 0$, the metric g_{ij} is unchanged if we shift the origin or rotate the coordinate system about the origin. This places a strong restriction on the form of the metric.

To understand the restriction, we should begin with Eq. (9.1) in spherical polar coordinates;

$$d\ell^2 = dx^2 + x^2\left(d\theta^2 + \sin\theta^2 d\phi^2\right). \tag{9.4}$$

The expression within the round bracket specifies the geometry on a sphere with unit radius. The geometry on the sphere is isotropic with respect to the origin $x = 0$, and the only generalisation of Eq. (9.4) that preserves the isotropy is of the form

$$d\ell^2 = A(x)x^2 + x^2\left(d\theta^2 + \sin^2\theta d\phi^2\right). \tag{9.5}$$

Homogeneity requires that the function $A(x)$ is unchanged if we change the origin of the coordinates. As I will now show, this requires $A = 1/(1 - Kx^2)$ for some constant K, giving

$$d\ell^2 = \frac{dx^2}{1 - Kx^2} + x^2\left(d\theta^2 + \sin^2\theta d\phi^2\right). \tag{9.6}$$

To show that homogeneity requires $A = 1/(1 - Kx^2)$ I will calculate the spatial curvature scalar and demand that it is independent of position as is required by the homogeneity of the geometry. The spatial curvature scalar is defined by the following sequence of formulas, which are like the formulas in Chapter 2 that define the spacetime curvature scalar.

$$\Gamma^m_{ij} = \frac{1}{2}g^{mn}\left(\frac{\partial g_{ni}}{\partial x^j} + \frac{\partial g_{nj}}{\partial x^i} - \frac{\partial g_{ij}}{\partial x^n}\right) \tag{9.7}$$

$$R^m_{nji} = \frac{\partial \Gamma^m_{nj}}{\partial x^i} - \frac{\partial \Gamma^m_{ni}}{\partial x^j} + \Gamma^m_{pi}\Gamma^p_{nj} - \Gamma^m_{pj}\Gamma^p_{ni} \tag{9.8}$$

$$R_{mn} = R^k_{mkn} \tag{9.9}$$

$$R = R^m_m. \tag{9.10}$$

Following the usual convention, I am denoting spatial curvature by the same letter R as spacetime curvature. The only difference between these formulas and the corresponding ones for spacetime is that the indices, denoted by Roman letters, run over 1 to 3 instead of 0 to 3.

Evaluating R with Eq. (9.5) one finds

$$R(x) = \frac{2}{x^2}\left[1 - \left(\frac{x}{A'(x)}\right)\right]. \tag{9.11}$$

Requiring $R = 6K$ for some constant K, this gives

$$A(x) = \frac{1}{1 - Kx^2 - C/x}, \tag{9.12}$$

where C is some constant. Since the metric is finite we need $C = 0$, which gives $A = 1/(1 - Kx^2)$ as required.

9.3 CLOSED GEOMETRY

We have seen that homogeneity and isotropy require a geometry of the form (9.6) for *some* K. Now I will show that a geometry of this form is homogeneous and isotropic for *every* K. Since $K = 0$ corresponds to the Euclidean case, we need only consider non-zero K. It is convenient to redefine the length scale through the changes $d\ell^2 \to d\ell^2/K$ and $x^2 \to x^2/K$. Then Eq. (9.6) becomes

$$d\ell^2 = \frac{dx^2}{1 \pm x^2} + x^2\left(d\theta^2 + \sin^2\theta d\phi^2\right) \tag{9.13}$$

where the sign is minus for positive K and positive for negative K. I deal in this section with positive K, which corresponds to the geometry proposed by Riemann.

For the minus sign we can write $x^2 = \sin^2\chi$. Then Eq. (9.13) becomes

$$d\ell^2 = d\chi^2 + \sin^2\chi\left(d\theta^2 + \sin^2\theta d\phi\right). \tag{9.14}$$

To show that this corresponds to a homogeneous and isotropic space, we need to go to a four-dimensional space with coordinates (x, y, z, w) which has Euclidean geometry defined by the line element

$$d\ell^2_{(4)} = dx^2 + dy^2 + dz^2 + dw^2. \tag{9.15}$$

This space is just a mathematical device, with no physical meaning. In it we can define a 3-sphere with radius r given by

$$x^2 + y^2 + z^2 + w^2 = r^2, \tag{9.16}$$

and on the 3-sphere we can define coordinates (θ, ϕ, χ) by

$$w = r\cos\chi \tag{9.17}$$
$$z = r\sin\chi\cos\theta \tag{9.18}$$
$$x = r\sin\chi\sin\theta\cos\phi \tag{9.19}$$
$$y = r\sin\chi\sin\theta\sin\phi. \tag{9.20}$$

With these coordinates Eq. (9.15) becomes

$$dl^2_{(4)} = r^2 dl^2 + dr^2,\qquad(9.21)$$

where dl^2 is given by Eq. (9.14). We see that the line element (9.13) with the minus sign is the metric of the 3-sphere.

To see that the metric of the 3-sphere is homogeneous and isotropic, note first that we could use as coordinates any three of (x, y, z, w), say (x, y, z).[2] To calculate dl^2 in terms of these coordinates, we should note first that Eq. (9.16) implies

$$x\,dx + y\,dy + z\,dz + w\,dw = 0.\qquad(9.22)$$

This gives an expression for dw on the sphere, and inserting it into Eq. (9.16) gives an expression for dl^2 in terms of dx, dy and dz. As this expression was obtained using Eqs. (9.16) and (9.21) which are invariant under rotations in the Euclidean space, we can use those rotations to shift the origin $x = y = z = 0$, or change the directions of the coordinate lines, without changing the expression for dl^2. This shows that the 3-sphere is indeed homogeneous and isotropic.

Since the space we deal with has the geometry of a 3-sphere, it has finite volume but no boundary. We can see this by looking at the line element defined by Eq. (9.14). According to Eq. (9.14) a sphere of radius χ space has area $4\pi \sin^2 \chi$, instead of the area $4\pi\chi^2$ that would apply to Euclidean space. As we increase χ the area of the sphere increases until $\chi = \pi/2$, but then it decreases and becomes zero when $\chi = \pi$. We have then covered the whole of the space, corresponding to the entire surface of the 3-sphere. The volume of the space is[3]

$$4\pi \int_0^\pi d\chi \sin^2 \chi = 2\pi.\qquad(9.23)$$

Because it is finite without a boundary, the Universe with this geometry is said to be closed.

9.4   OPEN GEOMETRY

Now we come to the case of negative K, which corresponds to the geometry of Bolyai and Lobachevsky. In this case the sign in Eq. (9.13) is positive. We can write $x^2 = \sinh^2 \chi$ to get

$$dl^2 = d\chi^2 + \sinh^2 \chi \left(d\theta^2 + \sin^2 \theta d\phi\right).\qquad(9.24)$$

This is the geometry proposed by Bolyai and Lobachevsky. The volume of a sphere with radius χ is now $4\pi \sinh^2 \chi$. It is bigger than the Euclidean volume $4\pi\chi^2$, and

[2]This is analogous to using coordinates (x, y) for the ordinary sphere $x^2 + y^2 + z^2 = r^2$. As in that case there is degeneracy, with pairs of points having the same coordinates, but that does not affect the proof of homogeneity and isotropy.

[3]This is with a particular choice for the distance unit, corresponding to the unit coefficient of $d\chi^2$ in Eq. (9.14). A different choice for the distance unit would correspond to a different coefficient.

it increases without limit as χ increases. The universe has no boundary, just like a Euclidean universe, and it is said to be open.

To show that the geometry is homogeneous and isotropic, we should use instead of Eq. (9.15) the line element for a Minkowski spacetime

$$ds^2 = dx^2 + dy^2 + dz^2 - dw^2, \qquad (9.25)$$

and work on the time-like slice defined by

$$x^2 + y^2 + z^2 - w^2 = -t^2, \qquad (9.26)$$

where t is a constant. (As before, this is just a mathematical device and the Minkowski spacetime that we are considering is not the one corresponding to physical spacetime.) On this slice we can define coordinates

$$w = t\cosh\chi \qquad (9.27)$$
$$z = t\sinh\chi\cos\theta \qquad (9.28)$$
$$x = = t\sinh\chi\sin\theta\cos\phi \qquad (9.29)$$
$$y = t\sinh\chi\sin\theta\sin\phi. \qquad (9.30)$$

With these coordinates, Eq. (9.25) becomes

$$ds^2 = t^2 d\ell^2 - dt^2, \qquad (9.31)$$

with now $d\ell^2$ given by Eq. (9.24). We see that $d\ell^2$ specifies the geometry of the Minkowski spacetime on the timelike slice.

The proof that the timelike slice is homogeneous and isotropic exactly mimics the proof that the 3-sphere is homogeneous and isotropic. We can use say (x, y, z) as coordinates, and we can use the Lorentz transformation to place the origin at any point on the timelike slice, and to orient the coordinate lines in any direction, without changing the metric when expressed in terms of these coordinates.

EXERCISES

1. Verify Eq. (9.11).

2. Verify Eq. (9.14).

3. For the geometry given by Eq. (9.14), calculate the circumference of a circle with radius r and compare it with the result for flat space.

4. Verify Eq. (9.21).

5. In Eqs. (9.7)–(9.10) one can allow the indices to run only over 1 and 2, so that one is dealing with a surface. In that case R is called the intrinsic curvature of the surface. Evaluate the intrinsic curvature of a sphere with radius r. Show that the intrinsic curvature of a cylinder vanishes. These results for the intrinsic curvature are in accordance with the fact that one can make a cylinder out of a flat sheet of paper, but not a sphere. Bearing that in mind, what is the only other kind of surface that has zero intrinsic curvature?

Newtonian perturbations

CONTENTS

I have so far taken the early universe to be completely homogeneous. It's time now to deal with the slight inhomogeneity, which causes the CMB anisotropy displayed in Figure 14.1 and leads eventually to the formation of galaxies. The quantities which define the slight departure from homogeneity are called **perturbations**. I will call the idealised, homogeneous Universe that we have been studying up till now, the **unperturbed Universe**.

In this chapter I deal with the perturbations using Newtonian physics, instead of General Relativity. Newtonian physics gives a good description of the matter after matter domination, within a region that is much smaller than the Hubble distance. The restriction on the size of the region is needed so that its edge recedes with speed much less than c.

10.1   UNPERTURBED UNIVERSE

Let us first ignore the perturbations, pretending that the Universe is perfectly homogeneous and isotropic. To describe the expansion, we take the cosmic fluid to be at rest at the origin of the position vector $\mathbf{r}$. Then the cosmic fluid at any other position has $\mathbf{r} \propto a(t)$, and it is useful to work with the comoving position vector $\mathbf{x}$ defined by $\mathbf{r}(t) = a(t)\mathbf{x}$. An object moving with the expansion has velocity

$$\mathbf{u}(t) = \dot{a}(t)\mathbf{x} = H(t)\mathbf{r}(t), \tag{10.1}$$

where $H = \dot{a}/a$ is the Hubble parameter. The mass continuity equation (3.32) in the unperturbed Universe is

$$\dot{\rho}(t) = -3H(t)\rho(t). \tag{10.2}$$

Since $H \equiv \dot{a}/a$, this corresponds to $a d\rho/da = -3\rho$ giving $\rho \propto a^{-3}$. That just says that the mass within an expanding volume is constant.

The potential at the edge of a sphere with radius $a(t)x$ coming from the mass within it is

$$\phi^{(0)}(\mathbf{x}, t) = \frac{-GM}{ax} = -\frac{4\pi G}{3}(ax)^2 \rho(t), \qquad (10.3)$$

where M is the mass within the sphere and ρ is the mass density. The potential coming from the mass between this sphere and a larger one is zero. If we take the second sphere to be big enough, the potential from the mass beyond it will be negligible which means that $\phi^{(0)}$ is the complete gravitational potential in the unperturbed Universe.

The acceleration of a fluid element with mass m is given by

$$m\frac{d\mathbf{u}}{dt} = -\boldsymbol{\nabla}\phi^{(0)}. \qquad (10.4)$$

The element has constant energy

$$E = m\phi^{(0)}(\mathbf{x}, t) + \frac{1}{2}m(\dot{a}x)^2 \qquad (10.5)$$

$$= -\frac{4\pi G}{3}m(ax)^2 \rho(t) + \frac{1}{2}m(\dot{a}x)^2. \qquad (10.6)$$

This is equivalent to Eq. (8.6) with $K = -(2E)/(mx^2)$.

The discussion so far invokes Newtonian gravity in its usual form, which has no place for the cosmological constant. We can include the effect of the cosmological constant on the expansion though, if we regard the Poisson equation (3.28) as the fundamental equation for determining the potential, rather than its Newtonian solution (3.29). Then, we can add to the Newtonian solution a term $\phi_\Lambda = \Lambda r^2/6$ where Λ is a constant. (We can do that because ϕ_Λ is a solution of $\nabla^2\phi_\Lambda = 0$.) Putting this term into the acceleration equation (3.27), we see that it adds to the inverse square law (3.27) a positive acceleration $\Lambda\mathbf{r}/3$. This is the acceleration that, according to Eq. (8.8), is generated in the relativistic theory by a cosmological constant ρ_Λ given by

$$8\pi G\rho_\Lambda = \Lambda. \qquad (10.7)$$

Taking $\mathbf{r} = 0$ to be our location, the positive acceleration is negligible within our galaxy, but at bigger distances it becomes significant and has to be included if the Newtonian theory is to correctly account for the effect of the cosmological constant on the evolution of the scale factor during matter domination.

10.2 TOTAL DENSITY PERTURBATION

The perturbation $\delta\rho$ in the total mass density is defined by

$$\rho(\mathbf{x}, t) = \rho(t) + \delta\rho(\mathbf{x}, t), \qquad (10.8)$$

where $\rho(t)$ is the density in some unperturbed Universe. The only essential requirement for the choice of $\rho(t)$ is that it makes $|\delta\rho| \ll \rho$. It is also convenient to consider the density contrast δ defined by

$$\delta \equiv \frac{\delta\rho(\mathbf{x},t)}{\rho(t)}. \tag{10.9}$$

The pressure perturbation is similarly defined:

$$P(\mathbf{x},t) = P(t) + \delta P(\mathbf{x},t). \tag{10.10}$$

We will also need a velocity perturbation, denoted by $\mathbf{v}$. This is defined by writing the velocity $\mathbf{u}$ of a fluid element with position $\mathbf{r}$ as

$$\mathbf{u}(\mathbf{x},t) = H(t)\mathbf{r}(\mathbf{x},t) + \mathbf{v}(\mathbf{x},t). \tag{10.11}$$

Finally, we need the perturbation in the potential, denoted by Φ:

$$\phi(\mathbf{x},t) = \phi^{(0)}(\mathbf{x},t) + \Phi(\mathbf{x},t), \tag{10.12}$$

where $\phi^{(0)}$ is the unperturbed potential.

We are going to work to first order in the perturbations. This means that we multiply all of the perturbations by some number ϵ, take the limit $\epsilon \to 0$, and then in the result set $\epsilon = 1$. Equivalently, it means that we only take the first term, in each of the infinite power series that would be needed for an exact treatment. The first order perturbations in the mass continuity, acceleration and Poisson equations give[1]

$$\dot{\delta} = -\boldsymbol{\nabla}\mathbf{v} \tag{10.13}$$

$$\dot{\mathbf{v}} + H\mathbf{v} + \boldsymbol{\nabla}\Phi = -\boldsymbol{\nabla}\delta P/\rho \tag{10.14}$$

$$\nabla^2\Phi = 4\pi G\rho\delta \tag{10.15}$$

$$= \frac{3}{2}H^2\delta. \tag{10.16}$$

In the final expression I used Eq. (8.4) to write ρ in terms of H.

The spatial average of a perturbation $g(\mathbf{x})$ can be removed by regarding it as part of the unperturbed quantity. That having being done, it is useful to work with the Fourier components, defined for each perturbation $g(\mathbf{x})$ by[2]

$$g(\mathbf{k}) = \int d^3\mathbf{x}\, g(\mathbf{x}) e^{-i\mathbf{k}\cdot\mathbf{x}}. \tag{10.17}$$

Since the physical distance is $\mathbf{r} = a(t)\mathbf{x}$, the physical wavenumber is $k/a(t)$. The inverse is

$$g(\mathbf{x}) = \frac{1}{(2\pi)^3} \int d^3\mathbf{k}\, g(\mathbf{k}) e^{i\mathbf{k}\cdot\mathbf{x}}. \tag{10.18}$$

---

[1]A dot can be taken to mean $\partial/\partial t$, though as we are working to first order it could as well be d/dt.

[2]Since the perturbations don't go to zero at infinity, this Fourier integral doesn't strictly speaking make sense. We should instead be dealing with a Fourier series, defined within a box with size $L \propto a$ that is now much bigger than the region of interest, and the integral (10.17) should go over this box.

Working with the Fourier components, Eqs. (10.13)–(10.16) become

$$\dot{\delta} = -i(\mathbf{k}/a) \cdot \mathbf{v} \tag{10.19}$$

$$\dot{\mathbf{v}} + H\mathbf{v} - i(\mathbf{k}/a)\Phi = -ik\delta P/\rho \tag{10.20}$$

$$-(k/a)^2\Phi(\mathbf{k}) = \frac{3}{2}H^2\delta(\mathbf{k}). \tag{10.21}$$

It is useful to write

$$\mathbf{v}(\mathbf{k}) = \mathbf{v}^{\mathrm{sc}} + \mathbf{v}^{\mathrm{vec}}, \tag{10.22}$$

where the first term (longitudinal or scalar contribution) is parallel to $\mathbf{k}$ and the second term (transverse or vector contribution) is perpendicular to it. The latter does not contribute to Eq. (10.19), and inserting it into Eq. (10.20) one finds

$$\dot{\mathbf{v}}^{\mathrm{vec}} + H\mathbf{v}^{\mathrm{vec}} = 0. \tag{10.23}$$

According to this equation, $\mathbf{v}^{\mathrm{vec}}$ decays like $1/a$.

It can be shown that the decay of $\mathbf{v}$ corresponds to the conservation of angular momentum within an expanding volume. Although we have derived this result using Newtonian physics, it holds also according to General Relativity.

To handle $\mathbf{v}^{\mathrm{sc}}$ it is useful to define a velocity potential V by writing

$$\mathbf{v}^{\mathrm{sc}}(\mathbf{k}) = -(i\mathbf{k}/k)V(\mathbf{k}). \tag{10.24}$$

Writing Eqs. (10.19) and (10.20) in terms of V one finds

$$\dot{\delta}(\mathbf{k}) = -(k/a)V(\mathbf{k}) \tag{10.25}$$

$$\dot{V}(\mathbf{k}) + HV(\mathbf{k}) - (k/a)\Phi(\mathbf{k}) = (k/a)\delta P/\rho. \tag{10.26}$$

In summary, we have found that the Newtonian perturbations fall into two classes, which evolve independently. These are the vector mode consisting of the single quantity $\mathbf{v}^{\mathrm{vec}}$ and the scalar mode consisting of $\mathbf{v}^{\mathrm{sc}}$, $\delta\rho$ and δP.

The pressure perturbation is negligible on sufficiently large wavelengths. Setting it to zero, we can eliminate V from Eqs. (10.25) and (10.26) and use Eq. (10.21), to get (suppressing the argument of $\delta(\mathbf{k})$)

$$\ddot{\delta} + 2H\dot{\delta} - \frac{3}{2}H^2\delta = 0. \tag{10.27}$$

As this equation doesn't involve $\mathbf{k}$, we can do the Fourier integral (10.18) to arrive at exactly the same equation for $\delta(\mathbf{k})$. The equation has a decaying solution proportional to t^{-1} and a growing solution proportional to $t^{2/3}$. We keep only the growing solution which according to Eq. (10.21) can be written

$$\delta(\mathbf{k}, t) = -\frac{2}{3}\left(\frac{k}{aH}\right)^2\Phi(\mathbf{k}), \tag{10.28}$$

with $\Phi(\mathbf{k})$ time independent. The time-dependence of δ can be written in three different ways:

$$\delta(\mathbf{x}, t) \propto t^{2/3} \propto a \propto 1/(aH)^2. \tag{10.29}$$

We shall see in Chapter 15 how the growth of δ leads to the formation of galaxies and galaxy clusters.

10.3 BARYONS AND CDM

As well as the total density contrast, we are interested in the separate density contrasts of the baryons and the CDM. Denoting them by subscripts b and c, we have

$$\dot{\delta}_c(\mathbf{k}) = -(k/a)V_c(\mathbf{k}) \tag{10.30}$$

$$\dot{V}_c(\mathbf{k}) + HV_c(\mathbf{k}) - (k/a)\Phi(\mathbf{k}) = 0 \tag{10.31}$$

$$\dot{\delta}_b(\mathbf{k}) = -(k/a)V_b(\mathbf{k}) \tag{10.32}$$

$$\dot{V}_b(\mathbf{k}) + HV_b(\mathbf{k}) - (k/a)\Phi(\mathbf{k}) = (k/a)\delta P_b/\rho_b. \tag{10.33}$$

Each species feels its own pressure gradient but the total gravitational acceleration.

Eliminating the velocities and using Eq. (10.21) for Φ, gives for the ordinary matter

$$\ddot{\delta}_b + 2H\dot{\delta}_b - \frac{3}{2}H^2\delta = -\left(\frac{k}{a}\right)^2 c_s^2 \delta_b, \tag{10.34}$$

where $c_s^2 \equiv \delta P_b/\delta\rho_b$. For the CDM it gives

$$\ddot{\delta}_c + 2H\dot{\delta}_c - \frac{3}{2}H^2\delta = 0. \tag{10.35}$$

The right hand side of Eq. (10.34) comes from the pressure perturbation. Let us suppose that it is negligible. Then $S_{cb} \equiv \delta_c - \delta_b$ satisfies

$$\ddot{S}_{cb} + 2H\dot{S}_{cb} = 0. \tag{10.36}$$

This has a constant solution, and a decaying solution $S \propto t^{-1/3}$. On the other hand, the total density contrast is

$$\delta = f_b\delta_b + f_c\delta_c \tag{10.37}$$

$$= \delta_b + f_c S_{cb} \tag{10.38}$$

$$= \delta_c - f_b S_{cb}, \tag{10.39}$$

where f_b is the fraction of the ordinary matter and f_c is the fraction of CDM.[3]

These equations first become valid at the start of the matter-dominated era. As we shall see, δ_b is then much smaller than δ_c so that $\delta \simeq f_c S_{cb}$ and $\delta_c \simeq S_{cb}$. Subsequently, δ grows and δ_c with it. As a result, the second terms of Eqs. (10.38) and (10.39) become negligible and we have

$$\delta \simeq \delta_b \simeq \delta_c. \tag{10.40}$$

There is now a common density contrast, applying to the ordinary matter, the CDM and the total.

---

[3]From Table 6.1 we see that $f_b = 0.18$ and $f_c = 0.82$, but the actual values are not important here.

This discussion applies if the pressure term in Eq. (10.34) is negligible. In the opposite case, the baryon density contrast oscillates with angular frequency $c_s k/a$. In contrast with its relativistic counterpart that we deal with in Section 13, this Newtonian oscillation is unimportant, and has not yet been observed.

The pressure term in Eq. (10.34) is negligible if the wavelength exceeds what is called the Jeans length, which is denoted by λ_J. It is useful to define also the Jeans mass M_J as the mass of matter within a sphere of radius $\lambda/2$. To estimate the Jeans length and mass, it is good enough to set $\delta_b = \delta$ in Eq. (10.34). Also, the three terms on the left hand side of Eq. (10.34) are roughly equal at the epoch $k = aH$ when the growth of δ can be said to begin. Equating the final term on the left hand side with the right hand side gives

$$\lambda_J = 2\pi c_s \left(4\pi G\rho\right)^{-1/2} \qquad c_s^2 \equiv \delta P_b/\delta \rho_b. \qquad (10.41)$$

For this expression to be useful, we need an estimate c_s. It can be shown that well before $z = 140$, the Compton scattering of photons off electrons keeps the baryon temperature T_b close to the photon temperature T_γ. Then, from the Maxwell-Boltzmann distribution, we have

$$c_s^2 = T_\gamma \frac{\delta n_b}{\delta \rho_b} = \frac{T_\gamma}{\mu m_p} \propto 1/a, \qquad (10.42)$$

where $\mu = 1.22$ is the mean molecular weight of the cosmic gas in atomic units and m_p is the proton mass. This gives a time-independent Jeans mass, $M_J = 1.4 \times 10^5 M_\odot$.

Going forward in time, one finds that at the epoch $z \simeq 10$ when the first galaxies form, $M_J = 6 \times 10^3 M_\odot$. As we see in Chapter 15, this provides an estimate of the mass of the lightest galaxies, which is in accordance with observation.

EXERCISES

1. Verify Eq. (10.2).

2. Use the chain rule to evaluate $\nabla^2 \phi(r)$ for a spherically symmetric potential. Verify that the only vacuum solutions of the Poisson equation in that case are $\phi \propto 1/r$ and $\phi \propto r^2$, as stated in the text.

3. An object with mass M generates a Newtonian potential $-GM/r$, while the cosmological constant contribution to the potential is $\Lambda r^2/6$. Use Eq. (10.7) and the value of ρ_Λ deduced from Table 6.1 to show that the Newtonian potential is bigger than the cosmological constant contribution if $r \lesssim (M/10^4 M_\odot)$ Mpc. Hence show that we can ignore the cosmological constant contribution when considering the motion of galaxies in the Newtonian regime $z \ll 1$.

4. Verify Eqs. (10.13)–(10.16).

5. Use Eq. (10.24) to show that $\boldsymbol{\nabla} \cdot \mathbf{v}(\mathbf{x}) = V(\mathbf{x})$ where $\boldsymbol{\nabla}$ has components $\partial/\partial x^i$. (This motivates the prefactor in Eq. (10.24).)

6. Verify Eq. (10.27).

7. Verify Eq. (10.36) and explain its significance.

8. Verify Eq. (10.42). What minimum galactic mass would it imply, if it continued to hold at the epoch of galaxy formation?

Relativistic perturbations

CONTENTS

Now we move beyond Newtonian physics, to consider cosmological perturbations in the context of General Relativity. We will be concerned with perturbations in the energy-momentum tensor and in the metric tensor.

In this chapter I deal with general properties of the perturbations. Then I describe what is called the primordial curvature perturbation, which as far as we know causes all of the other perturbations.

11.1   DEFINING THE PERTURBATIONS

To define the relativistic perturbations, we have to make a choice for the coordinates (x, y, z, t) that describe the perturbed Universe. The lines of fixed (x, y, z) are called threads of the Universe, and the three-dimensional regions of spacetime with fixed t are called slices.

In the simplest case, we deal with the perturbation of a quantity f, which is nonzero in the unperturbed Universe and is specified by a single number. Then, as in Eq. (10.8), the perturbation δf is defined by

$$f(\mathbf{x}, t) = f(t) + \delta f(\mathbf{x}, t), \tag{11.1}$$

where $f(t)$ refers to some unperturbed Universe.

To deal with the present accuracy of the observations, we can use the linear approximation, i.e. work to first order in the perturbations. This gives a simple expression for the effect of a change of coordinates on a perturbation. After changing to new coordinates $\mathbf{x}'(\mathbf{x}, t)$ and $t'(\mathbf{x}, t)$, we have instead of Eq. (11.1)

$$f(\mathbf{x}', t') = f(t') + \delta f(\mathbf{x}', t'). \tag{11.2}$$

At a given point in spacetime, the new and old coordinates have different numerical values. Taking instead the numerical values to be the same, the new and old

time

position

Figure 11.1   The dashed line represents a slice of spacetime corresponding to some coordinate choice $(\mathbf{x}, t)$, and the solid line represents a slice corresponding to a different coordinate choice $\mathbf{x}'(\mathbf{x}, t)$ and $t'(\mathbf{x}, t)$ with the same numerical values for the coordinates.

coordinates refer to different spacetime points. As shown in Figure 11.1 they then define different slicings and threadings. With the same numerical values, the first terms of Eqs. (11.1) and (11.2) are the same and we have

$$\delta f(\mathbf{x}', t') - \delta f(\mathbf{x}, t) = f'(\mathbf{x}', t') - f(\mathbf{x}, t). \tag{11.3}$$

The change in coordinates should be regarded as a perturbation, so that it does not generate a large inhomogeneity. Then, the change in the threading is a second order effect, because it affects only the perturbation (not the unperturbed quantity because that is independent of position). Ignoring the change in threading, the change in coordinates just shifts the time corresponding to equal numerical values of the coordinates, by some amount $\delta t(\mathbf{x}, t)$. To first order this changes the perturbation by an amount

$$\delta f'(\mathbf{x}', t') - \delta f(\mathbf{x}, t) = \frac{\partial f(\mathbf{x}, t)}{\partial t} \delta t(\mathbf{x}, t), \tag{11.4}$$

but we can here ignore the difference between $f(\mathbf{x}, t)$ and $f(t)$ because it is a second-order effect, which gives finally

$$\delta f'(\mathbf{x}', t') - \delta f(\mathbf{x}, t) = \dot{f}(t)\delta t(\mathbf{x}, t). \tag{11.5}$$

As in the Newtonian case, it is useful to work with the Fourier components $g(\mathbf{k})$ of a perturbation $g(\mathbf{x})$, defined by Eq. (10.17). The wavelength is $2\pi a(t)/k$, and I will call $1/k$ the **scale** of the Fourier component. We are interested only in scales that can be probed by cosmological observations, which I will call **cosmological scales**. The biggest cosmological scale is roughly the present size of the observable Universe, $\sim 10^4$ Mpc. The smallest cosmological scale is the one that leads to the formation of the smallest galaxies, which as we see in Chapter 15 is about 10^{-3} Mpc.

At the beginning of the know history all cosmological scales have $aH/k \gg 1$ corresponding to wavelengths much bigger than the Hubble distance. As time goes on, successively smaller scales reach the epoch when $aH/k = 1$. I will call that the epoch of horizon entry because a/k is roughly the wavelength and $1/H$ is about equal to the distance to the horizon. Epochs of horizon entry for some cosmological scales are shown in the second column of Table 11.1. The third and fourth columns of that table will be explained later.

Scale $1/k$ (Mpc)	z when $k = aH$	CMB ℓ	mass of galaxy
1.0×10^{-3}	3.3×10^8		$10^3 M_\odot$
1.0×10^{-2}	3.3×10^7		$10^6 M_\odot$
0.10	3.3×10^6		$10^9 M_\odot$
1.0	3.3×10^5		$10^{12} M_\odot$
10	3.3×10^4	1400	$10^{15} M_\odot$
89	3.3×10^3	157	
2.1×10^2	1050	63	
1.0×10^3	94	14	

Table 11.1 **Cosmological scales** The first column gives the scale $1/k$, for some scales of cosmological interest. The second column gives the redshift $z \simeq 1/a$ at the epoch when $a/k = 1/H$, which is roughly the epoch of horizon entry. The smallest cosmological scale enters the horizon just after electron-positron annihilation, and bigger scales enter the horizon later. Horizon entry occurs during matter domination for rows below the line, and during radiation domination for rows above it. Last scattering occurs at $z = 1100$. The third column gives the approximate scale of the CMB that is explored by a given k, which is $\ell \simeq k x_{ls}$. The fourth column gives the mass of the object formed when a region within the smoothing radius collapses, calculated in Chapter 15.

11.2 STATISTICAL PROPERTIES

We are not interested in the details of the perturbations. In particular, we are not interested in the locations of the individual peaks of the CDM's smoothed density contrast, that will become galaxies. The locations of individual galaxies are accidents, just as the distance of the Earth from the sun is an accident.

We are interested instead, in the *statistical properties* of the perturbations. To understand what that means, let's imagine that a coin is tossed many times, and the results are written down. Then we could highlight all the 'heads' entries, and see how they are distributed. On average there would be as many heads as tails, but the distribution would be uneven and we would sometimes find two heads together. We might find even three or more but that becomes increasingly unlikely. We could work out of the probability for each of these possibilities, and of any other possibility regarding the distribution of 'heads'. Then we would be studying the statistical properties of the distribution of 'heads'.

The statistical properties of the perturbations refer to their form at a given epoch. To define them, one must in general invoke an ensemble of Universes, of which ours is supposed to be a typical member. As we will see in Chapter 19, such an ensemble is provided by quantum physics, if the perturbations originate as a quantum fluctuation during inflation. Given an ensemble, the statistical properties of a perturbation $g(\mathbf{x})$ are defined by giving $\langle g(\mathbf{x}_1)g(\mathbf{x}_2)\rangle$, $\langle g(\mathbf{x}_1)g(\mathbf{x}_2)g(\mathbf{x}_3)\rangle$ and so on, where brackets denote averages over the ensemble. These are called correlators.

As is implied by the notation, one uses the comoving position $\mathbf{x} = \mathbf{r}/a$ to specify the correlators.

The correlators are usually supposed to be invariant under translations (statistical homogeneity) and under rotations (statistical isotropy). Assuming statistical homogeneity, I show at the end of this section that the bracket defining each correlator can be regarded as the average over space, with the differences $\mathbf{x}_1 - \mathbf{x}_2$ etc. fixed. This is called the ergodic theorem, and it means that we can define the correlators using a single realization of the ensemble which we of course take to be the actual Universe.

Let us regard $\langle\rangle$ as an ensemble average. Then translation invariance requires $\langle g(\mathbf{k})\rangle = 0$ and it determines the form of $\langle g(\mathbf{k})g(\mathbf{k}')\rangle$. We can write

$$\langle g(\mathbf{k})g(\mathbf{k}')\rangle = \int d^3x d^3x' \langle g(\mathbf{x})g(\mathbf{x}+\mathbf{x}')\rangle e^{i[\mathbf{k}\cdot\mathbf{x}+\mathbf{k}'\cdot(\mathbf{x}+\mathbf{x}')]}. \tag{11.6}$$

The translation invariance means that $\langle g(\mathbf{x})g(\mathbf{x}+\mathbf{x}')\rangle$ is independent of $\mathbf{x}$. Using

$$\int e^{i(\mathbf{k}+\mathbf{k}')\cdot\mathbf{x}} d^3x = (2\pi)^3 \delta^2(\mathbf{k}+\mathbf{k}') \tag{11.7}$$

that gives

$$\langle g(\mathbf{k})g(\mathbf{k}')\rangle = (2\pi)^3 \delta^3(\mathbf{k}+\mathbf{k}') P_g(k),, \tag{11.8}$$

where

$$P_g(\mathbf{k}') = \int d^3x' \langle g(\mathbf{x})g(\mathbf{x}+\mathbf{x}')\rangle e^{i\mathbf{k}'\cdot\mathbf{x}'}. \tag{11.9}$$

Translation invariance allows us to set $\mathbf{k}$ to 0. Also, P_g depends only on k' because of rotational invariance. Dropping the prime on k' we can therefore write

$$P_g(k) = \int d^3x \langle g(0)g(\mathbf{x})\rangle e^{i\mathbf{k}\cdot\mathbf{x}}. \tag{11.10}$$

Using Eqs. (11.7) and (11.8), the mean square $\sigma_g^2 \equiv \langle g^2(\mathbf{r})\rangle$ is seen to be

$$\sigma_g^2 = \int_0^\infty \mathcal{P}_g(k) dk/k \tag{11.11}$$

where $\mathcal{P}_g(k) \equiv (k^3/2\pi^2)P_g(k)$ is called the spectrum of g. Using Eq. (11.8) it is given by

$$\langle g(\mathbf{k})g(\mathbf{k}')\rangle = (2\pi)^3 \frac{2\pi^2}{k^3} \delta^3(\mathbf{k}+\mathbf{k}') \mathcal{P}_g(k),. \tag{11.12}$$

From now on I will use $\mathcal{P}_g$ instead of P_g.

The two-point correlator (11.8) cannot be zero, because the reality condition $g(-\mathbf{k}) = g^*(\mathbf{x})$ means that it is specifying $\langle|g(\mathbf{k})|^2\rangle$. It's generally assumed that at early times, when the perturbations are sufficiently small and the linear approximation applies, the reality condition provides the only correlation between the Fourier components. This property of the Fourier components is called Gaussianity, and is consistent with observation at the present level of accuracy.

To see why the property is called Gaussianity, we should consider the real and imaginary parts of the Fourier components. The Gaussianity property means that the real and imaginary parts are completely uncorrelated, implying that $g(\mathbf{x})$ is an infinite sum of uncorrelated quantities. According to a fundamental result of statistics, such a sum always leads to a Gaussian probability distribution.[1]

For a Gaussian perturbation, the correlator involving any number of perturbations can be calculated by considering all possible pairs of the perturbations, and then replacing one of the pair by the complex conjugate $g*(-\mathbf{k}) = g(\mathbf{k})$. For four perturbations this gives

$$
\begin{aligned}
\langle g(\mathbf{k}_1)g(\mathbf{k}_2)g(\mathbf{k}_3)g(\mathbf{k}_4)\rangle &= \langle g(\mathbf{k}_1)g(\mathbf{k}_2)\rangle\langle g(\mathbf{k}_3)g(\mathbf{k}_4)\rangle + \langle g(\mathbf{k}_1)g(\mathbf{k}_3)\rangle\langle g(\mathbf{k}_4)g(\mathbf{k}_2)\rangle \\
&+ \langle g(\mathbf{k}_1)g(\mathbf{k}_4)\rangle\langle g(\mathbf{k}_3)g(\mathbf{k}_2)\rangle,
\end{aligned}
\tag{11.13}
$$

and a similar expression holds for any even number. For an odd number, the same procedure can be used to deal with all except one of the perturbations, but the expectation value of that one is zero which means that the whole correlator is zero.

To prove the ergodic theorem, we should consider a finite box. Regarding $\langle\rangle$ as a spatial average we then have[2]

$$
\begin{aligned}
\langle g(\mathbf{x})g(\mathbf{x}+\mathbf{y})\rangle &= L^{-3}\int g(\mathbf{x})g(\mathbf{x}+\mathbf{y})d^3y \tag{11.14} \\
&= L^{-3}(2\pi)^{-6}\int g(\mathbf{k})g(\mathbf{k}')e^{i[\mathbf{k}\cdot\mathbf{y}+\mathbf{k}'\cdot(\mathbf{x}+\mathbf{y})]}d^3y\,d^3k\,d^3k' \tag{11.15} \\
&= L^{-3}(2\pi)^{-3}\int \delta*3(\mathbf{k}+\mathbf{k}')g(\mathbf{k})g(\mathbf{k}')e^{i\mathbf{k}\cdot\mathbf{x}}d^3k\,d^3k'. \tag{11.16}
\end{aligned}
$$

Within each element d^3k, translation invariance demands that the Fourier components are uncorrelated.[3] We can therefore replace $g(\mathbf{k})g(\mathbf{k}')$ by $\langle g(\mathbf{k})g(\mathbf{k}')\rangle$ which gives

$$
\begin{aligned}
\langle g(\mathbf{x})g(\mathbf{x}+\mathbf{y})\rangle &= L^{-3}(2\pi)^{-6}\int \delta^3(\mathbf{k}+\mathbf{k}')\langle g(\mathbf{k})g(\mathbf{k}')\rangle e^{i\mathbf{k}\cdot\mathbf{x}}d^3k\,d^3k' \tag{11.17} \\
&= L^{-3}\int [\delta^3(\mathbf{k}+\mathbf{k}')]^2 \mathcal{P}_g(k)e^{i\mathbf{k}\cdot\mathbf{x}}d^3k\,d^3k' \tag{11.18} \\
&= \left(\frac{1}{2\pi}\right)^3\int \mathcal{P}_g(k)e^{i\mathbf{k}\cdot\mathbf{x}}d^3k. \tag{11.19}
\end{aligned}
$$

---

[1] For example, if you toss a coin many times, giving yourself a score of $+1$ for heads and -1 for tails, the probability distribution of your total score will be Gaussian.

[2] The Fourier integral is used as a good approximation to the Fourier series, which is valid because the box is supposed to be bigger than the region of interest.

[3] For a Gaussian perturbation this is a consequence of the delta function in Eq. (11.8). For a non-Gaussian perturbation, higher correlators are nonzero but translation invariance still gives delta functions which ensure that the Fourier components within a cell are uncorrelated.

(To get the last line I used the expression

$$[\delta^3(\mathbf{k} - \mathbf{k}')]^2 = (L/2\pi)^3 \delta(\mathbf{k} - \mathbf{k}'), \tag{11.20}$$

which is justified by the correspondence

$$\left(\frac{2\pi}{L}\right)^3 \sum_n \longrightarrow \int d^3k \tag{11.21}$$

that takes us from the Fourier series to the Fourier integral.) Eq. (11.19) is a Fourier integral as in Eq. (10.18), whose inverse as in Eq. (10.17) coincides with Eq. (11.8). This completes the proof of the ergodic theorem.

11.3 SMOOTHING

Instead of the actual Universe, it is often convenient to consider a **smoothed** Universe. The smoothed Universe can be thought of as the one that an outsider would see, if they had poor eyesight. To be precise, the value of a quantity in the smoothed Universe at a given point in space is the average of the actual quantity within a sphere centred at that point.

I will take the radius of the smoothing sphere to be proportional to the scale factor a, and call it the **smoothing radius**. Then I write

$$\text{smoothing radius} = Ra(t), \tag{11.22}$$

and will call R the **smoothing scale**. Smoothing multiplies each Fourier component by a window function $W(kR)$, given by

$$W(kR) = \frac{3}{4\pi R^3} \int_{|\mathbf{x}|<R} d^3x\, e^{i\mathbf{k}\cdot\mathbf{x}} \tag{11.23}$$

$$= 3\left[\frac{\sin(kR)}{(kR)^3} - \frac{\cos(kR)}{(kR)^2}\right]. \tag{11.24}$$

The window function is equal to 1 at $k = 0$ and is much less than 1 at $k \gg R^{-1}$. Roughly speaking, the effect of smoothing is to cut out components with $k \gtrsim R$ while leaving unaltered components with $k \lesssim R^{-1}$.

The smoothed Universe is practically homogeneous within a region that is much smaller than the smoothing radius. If we take the smoothing scale to be much bigger than the present Hubble distance c/H_0, we arrive at the (practically) homogeneous observable Universe that was the subject of earlier chapters. But there's nothing special about the present — the Universe doesn't know we're here! If we go back in time, we can arrive at an era when the smoothing radius is much bigger than the Hubble distance c/H even if it is not so at present.[4] Then, the Universe at each

---

[4]The smoothing radius Ra decreases more slowly than the Hubble distance $ca/\dot{a}$ as we go back in time, because gravity makes $\ddot{a}$ negative.

location looks like some homogeneous Universe. As a result, *when the smoothing radius is much bigger than the Hubble distance, each quantity at a given location evolves as it would in some homogeneous Universe.* In effect, the whole Universe becomes a collection of separate homogeneous Universes. This is usually called the **separate Universe assumption**, though it is actually an inevitable consequence of the fact that the Hubble distance is the biggest relevant distance at each epoch.

The range of Fourier scales $1/k$ that is of cosmological interest was shown in Table 11.1. Since smoothing with smoothing scale R cuts out Fourier components with $k \lesssim R^{-1}$, the range of R of cosmological interest is the same as the range of $1/k$. At the beginning of the known history, the smoothing radius aR is bigger than the Hubble distance for all R of cosmological interest.

11.4   PRIMORDIAL CURVATURE PERTURBATION

As far as we can tell, all perturbations of the cosmic fluid on cosmological scales are caused by a single perturbation that exists at the beginning of the known history. I will call this perturbation the **primordial curvature perturbation**, and the beginning of the known history the primordial epoch.

To define the primordial curvature perturbation, one should smooth the Universe with smoothing radius that is much bigger than the Hubble distance at the primordial epoch, but smaller than the smallest cosmological scale $1/k$. (Such a choice is possible, as we saw at the end of the previous section and as is shown in Table 11.1.) Then one should choose the slices of spacetime to have homogeneous energy density, and choose the threads to be comoving (moving with the expansion). Ignoring gravitational waves, the spatial metric can then be written[5]

$$g_{ij} = a^2(\mathbf{x}, t)\delta_{ij}. \tag{11.25}$$

The local scale factor $a(\mathbf{x}, t)$ determines the rate of increase in the volume of an element of the cosmic fluid. The primordial curvature perturbation ζ is defined by writing[6]

$$a(\mathbf{x}, t) = a(t)e^{\zeta(\mathbf{x}, t)}. \tag{11.26}$$

The primordial curvature perturbation is defined on a slice of uniform total energy density. The **adiabatic condition** states that the energy density of each of the four components is also uniform on this slice.[7] In other words, these separate energy densities are determined by the total energy density. The adiabatic condition is consistent with observation at the present level of accuracy.

I will now show that the adiabatic condition makes ζ constant. To do that I will use the separate Universe assumption. As the smoothing radius is bigger than the

---

[5]Gravitational waves replace δ_{ij} by a function that distorts the shape of the element without affecting its volume, which is irrelevant in this context.

[6]It's called the curvature perturbation because it precludes the possibility of finding Cartesian coordinates, such that $ds^2 = a^2(t)(x^2 + dy^2 + dz^2)$.

[7]These are the baryons, the CDM, the photons and the neutrinos.

Hubble distance, the Universe looks homogeneous at each location so that Eq. (6.6) applies. That equation gives, for any choice of the time coordinate t,

$$\dot{\rho} = -3(\dot{a}/a)(\rho + P). \tag{11.27}$$

According to the adiabatic condition, the pressure is homogeneous on the slices of homogeneous energy density. Using Eq. (11.26) we therefore have on the slicing of uniform ρ,

$$\dot{\rho}(t) = -3\left[\frac{\dot{a}(t)}{a(t)} + \dot{\zeta}(\mathbf{x},t)\right](\rho(t) + P(t)). \tag{11.28}$$

This makes the rate of change of ζ independent of position. Choosing the spatial average of ζ to be zero, which is possible because a spatially homogeneous ζ could be regarded as belonging to $a(t)$, we conclude that ζ is indeed constant.

As ζ is supposed to be the cause of all subsequent perturbations, it can be inferred from observation. We are interested in the spectrum $\mathcal{P}_\zeta(k)$, and a good fit to observation is provided by

$$\mathcal{P}_\zeta(k) = A(k/k_0)^{n_s-1} \tag{11.29}$$

with $k_0 = 0.05\,\mathrm{Mpc}^{-1}$ and the values of A and n_s that are shown in Table A.3. The scale $1/k_0$, called the pivot scale, is chosen to be roughly in the middle of the logarithmic range of scales probed by observation.

Within the accuracy of current observations, it is good enough to make the linear approximation $e^\zeta = 1 + \zeta$. This gives

$$\zeta = \frac{\delta a}{a} = \frac{1}{2}\frac{\delta(a^2)}{a^2}, \tag{11.30}$$

which I will use from now on. It corresponds to the spatial metric

$$g_{ij} = a^2(\eta)\delta_{ij}(1 + 2\zeta). \tag{11.31}$$

Using the linear approximation, the adiabatic condition relates ζ to the primordial energy density perturbations $\delta\rho_i$ of the components of the cosmic fluid. It is convenient to consider the smoothed density *contrasts*, $\delta_i \equiv \delta\rho_i/\rho$. According to the adiabatic condition, the smoothed energy density of each component of the cosmic fluid is homogeneous on the slicing that makes the total smoothed energy density homogeneous. To get nonzero energy density perturbations, we need to go to a different slicing with a time shift $\delta t(\mathbf{x},t)$. According to Eq. (11.5), this gives for a radiation component

$$\delta_r = \frac{\dot{\rho}_r}{\rho_r}\delta t = -3H\frac{\rho_r + P_r}{\rho_r}\delta t = -4H\delta t. \tag{11.32}$$

For a matter component the pressure is negligible which gives instead $\delta_m = -3H\delta t$. We conclude that the radiation components have a common smoothed density contrast, and so do the matter components, the two being related by

$$\delta_r = \frac{3}{4}\delta_m. \tag{11.33}$$

EXERCISES

1. Explain how the ergodic theorem simplifies the description of cosmological perturbations. Under what circumstance would the ergodic theorem be invalid? How would that alter the description of cosmological perturbations?

2. Verify Eq. (11.8).

3. Write down the version of Eq. (11.11), which would give σ_g^2 in terms of a $P_g(\mathbf{k})$ allowing that quantity to depend on direction.

4. Write Eq. (11.23) in terms of a double integral over spherical polar coordinates whose pole points along $\mathbf{k}$. Evaluate the integral over θ. Then verify that the integral over R gives Eq. (11.24), by differentiating that expression with respect to R.

5. The justification given in footnote 4, of the statement in the text, ignores the cosmological constant. Add a sentence to the justification to take care of that omission.

6. In general, P in Eq. (11.28) could depend on position and then ζ would vary with time. Working to first order, find the expression relating $\dot{\zeta}$ to δP.

CHAPTER 12

Scalar perturbations

CONTENTS

12.1   TENSOR, VECTOR AND SCALAR MODES

We saw in Chapter 10 that Newtonian perturbations consist of two modes, which evolve independently. These are the vector mode which describes the vorticity of the fluid flow, and the scalar mode which among other things describes the perturbations in the energy density. For relativistic perturbations there is in addition a tensor mode, which describes gravitational waves and their interaction with the cosmic fluid.

The three modes are defined by the behaviour of the Fourier components of the perturbations, under a rotation about the **k** direction. Scalar perturbations are not changed by any rotation. Vector perturbations are unchanged only if we rotate by 2π.[1] Tensor perturbations are unchanged if we rotate just by π.[2] Since the form of the equations must be unaffected by a rotation, the three modes evolve independently.

The gravitational waves that correspond to the tensor mode were dealt with in Sections 3.6 and 4.1. As we will see in Chapter 18, the tensor mode is generated by inflation at some level, though it may be too small ever to observe. The vector mode is assumed absent because it would decay with time. That leaves only the scalar mode, which is our focus for the rest of the chapter.

12.2   SCALAR PERTURBATIONS

To make the text easier to follow I will make some statements without their proof, which is given in Appendix A.2. Allowing for all three modes, the perturbed metric

---

[1]This is clearly so in the Newtonian case, where the only vector perturbation is the transverse component of the fluid velocity.

[2]For gravitational waves, this is clear from the discussion in Section 3.6.

tensor corresponds to a line element of the form

$$ds^2 = a^2(\eta) \left[-(1+2A)\,d\eta^2 - 2B_i d\eta dx^i + (\delta_{ij}(1+2D) + 2E_{ij})\,dx^i dx^j \right], \quad (12.1)$$

where E_{ij} is taken to be traceless. For the scalar mode, the Fourier components of B_i and E_{ij} can be written

$$B_i(\mathbf{k}) = -\frac{ik_i}{k} B(\mathbf{k}) \quad\quad (12.2)$$

$$E_{ij}(\mathbf{k}) = \left(-\frac{k_i k_j}{k^2} + \frac{1}{3}\delta_{ij} \right) E(\mathbf{k}). \quad\quad (12.3)$$

The scalar perturbations of the metric tensor are therefore, in general, defined by A, B, D and E.

There is a unique choice for the coordinates which makes $B = E = 0$. The coordinate system with that choice is called the longitudinal gauge and I will use it to describe the perturbations. Adopting the usual notation $A = \Psi$ and $D = -\Phi$, the line element is

$$ds^2 = a^2(\eta) \left[-(1+2\Psi)\,d\eta^2 + (1-2\Phi)\,\delta_{ij}dx^i dx^j \right]. \quad\quad (12.4)$$

We also need to consider the perturbation of the energy-momentum tensor. As seen in Section 3.1 its components are[3]

$$\delta T^{00} = \delta\rho \quad\quad (12.5)$$

$$\delta T^{0i} = \delta T^{i0} = v^i(\rho + P) \qu\quad (12.6)$$

$$\delta T^{ij} = \delta_{ij}\delta P + \Sigma_{ij}. \qu\quad (12.7)$$

In the scalar mode we can write

$$v^i = -i\frac{k^i}{k} V \qu\quad (12.8)$$

$$\Sigma_{ij} = \left(-\frac{k_i k_j}{k^2} + \frac{1}{3}\delta_{ij} \right) P\Pi. \qu\quad (12.9)$$

The scalar perturbations of the energy-momentum tensor are therefore defined by $\delta\rho$, δP, V and Π.

12.3 EVOLUTION OF THE SCALAR PERTURBATIONS

We need equations for the time dependence of the perturbations. Instead of d/dt it is more convenient to use conformal time η defined by $d/d\eta = a\,d/dt$, because we will find oscillations whose frequency with respect to η is constant or slowly

---

[3]I am not attaching a δ to the perturbations v^i and Σ_{ij} because they vanish in the unperturbed Universe.

varying. *Therefore an overdot in this chapter and the next will denote $d/d\eta$.* I work with Fourier components but suppress the argument **k**.

The perturbation of the energy continuity equation $D_\mu T^{\mu 0}$ gives

$$\dot{\delta} = -(1+w)\left(kV - 3\dot{\Phi}\right) + 3aHw\left(\delta - \frac{\delta P}{P}\right), \qquad (12.10)$$

where $w \equiv P/\rho$. The perturbation of the acceleration equation $D_\mu T^{\mu i}$ gives

$$\dot{V} = -aH(1-3w)V - \frac{\dot{w}}{1+w}V + k\frac{\delta P}{\rho + P} - \frac{2}{3}k\frac{w}{1+w}\Pi + k\Psi. \qquad (12.11)$$

The Einstein field equation gives two more expressions

$$\delta + 3\frac{aH}{k}(1+w)V = -\frac{2}{3}\left(\frac{k}{aH}\right)^2 \Phi \qquad (12.12)$$

and

$$\Pi = \left(\frac{k}{aH}\right)^2 (\Psi - \Phi).. \qquad (12.13)$$

It will also be useful to invoke in addition the following relation, valid in the limit $k \to 0$.[4]

$$\delta = -2\left[\Psi + (aH)^{-1}\dot{\Phi}\right].. \qquad (12.14)$$

Well after horizon exit one expects that Eq. (12.12) becomes

$$\delta = -\frac{2}{3}\left(\frac{k}{aH}\right)^2 \Phi. \qquad (12.15)$$

This expectation can be verified after neutrino decoupling, for the growing mode which is the one that matters.

We will also need to consider the perturbations of individual components of the cosmic fluid. The total perturbations in the energy density and pressure are the sums of the individual perturbations. From Eq. (3.2), the total velocity potential is given in terms of the individual velocity potentials by the relation by

$$(\rho + P)V = \sum_i (\rho_i + P_i)V_i. \qquad (12.16)$$

---

[4]This corresponds to the $k = 0$ limit of the time-time component of the Einstein equation. It can be derived from Eqs. (12.10)–(12.13).

12.4 INITIAL CONDITION

On cosmological scales, we need to follow the evolution of the perturbations described in the previous section. To do that, we need initial conditions laid down at the beginning of the known history. Given the adiabatic condition, it can be shown that the initial conditions are entirely determined by the constant value of the primordial curvature perturbation $\zeta(\mathbf{k})$. To simplify the discussion I will ignore the neutrinos. Then thermal equilibrium ensures that the cosmic fluid is isotropic in its rest frame, which means that the anisotropic stress vanishes giving $\Psi = \Phi$.

We need to relate $-\Phi$, defined by Eq. (12.4), to ζ defined by Eq. (11.31). Both of them represent a fraction change $\delta a/a$ in the scale factor. Since $\delta a/a = \delta(\ln a)$, it follows from Eq. (11.5) that

$$\zeta + \Phi = \frac{d\ln a}{dt}\delta t = H\delta t, \tag{12.17}$$

where δt is the time shift going from the slicing on which $-\Phi$ is defined, to the slicing of uniform energy density on which ζ is defined. We are defining $\delta\rho$ on the former slicing which means that

$$\delta\rho = -\frac{d\rho}{dt}\delta t = 3H(\rho + P)\delta t. \tag{12.18}$$

Putting this into Eq. (12.17) we get

$$\zeta + \Phi = \frac{\delta}{3(1+w)}. \tag{12.19}$$

Then Eq. (12.14) gives, after setting $\Phi = \Psi$ and dropping a decaying solution,

$$\Phi = \Psi = -\frac{3+3w}{5+3w}\zeta, \tag{12.20}$$

with $w = 1/3$ because we deal with a radiation-dominated era. With Eq. (12.19) this gives

$$\delta_\gamma = -2\Phi = \frac{4}{3}\zeta, \tag{12.21}$$

where I have set $\delta = \delta_\gamma$ because we are ignoring the neutrinos. The adiabatic condition (11.33) gives for the matter density contrasts

$$\delta_c = \delta_b = \frac{4}{3}\delta_\gamma.. \tag{12.22}$$

EXERCISES

1. Show that the E_{ij} defined by Eq. (12.3) is traceless and invariant under rotations about the $\mathbf{k}$ direction. Explain why it is the most general matrix with these properties.

2. Write down the components of a 4-vector A^μ which lies along the threads defined by Eq. (12.4), and the components of a 4-vector B^μ which lies on a slice (both at the same spacetime position). Verify that $g_{\mu\nu} A^\mu B^\nu$ vanishes.

3. Verify that Eqs. (12.19) and (12.20) give Eq. (12.21).

CHAPTER 13

Baryon acoustic oscillation

CONTENTS

13.1   GROWTH, OSCILLATION AND DECAY

While the smoothing scale is outside the horizon, there is no time for anything to move appreciably. As seen in Table 11.1, the smoothing scale for cosmological scales is outside the horizon at the epoch just after electron-positron annihilation. We can therefore begin our discussion at that epoch, when the cosmic fluid has just four components. These are the photons, neutrinos, CDM and baryons.

The smoothing scale enters the horizon on successively smaller cosmological scales, as shown in Table 11.1. After horizon entry each component of the cosmic fluid is free to move. There is then a competition between the random motion of the particles, which tries to push them apart, and gravity which tries to pull things together. The result of this competition is described by what are called the Boltzmann equations. In this book I will not deal with the Boltzmann equations, though I will give results that are obtained by using them. Instead, I will make approximations that give a feel for the essential physics.

Before getting into detail, it will be useful to summarise the behaviour of each component of the cosmic fluid. To do that, it is enough to focus on the energy density contrast $\delta \equiv \delta\rho/\rho$.

For the CDM, random motion is negligible and gravity wins.[1] As a result, the CDM density contrast grows after horizon entry. As we will see, the growth is only logarithmic during radiation domination, but then becomes more rapid with the density contrast proportional to a. For neutrinos the random motion is relativistic at horizon entry. The neutrinos travel freely and their density contrast falls exponentially.

For baryons and photons, the behaviour is more complicated. Let us first consider wavelengths that enter the horizon before last scattering. Then the scattering of photons off electrons inhibits their diffusion and we deal with a single fluid called

---

[1]This is part of the definition of CDM. One can consider more complicated scenarios which extend the definition of CDM to allow it to have significant random motion. It is then called Warm Dark Matter.

the baryon-photon fluid. The fluid has high pressure, coming from the motion of the photons. As a result, the pressure of the fluid undergoes what is called the **baryon acoustic oscillation**. The amplitude of the baryon acoustic oscillation undergoes what is called Silk damping, caused by photon diffusion.

After last scattering, the photons travel freely, to become the CMB that will be discussed in Chapter 14. The baryon pressure now becomes negligible on scales above what is called the Jeans scale, allowing the baryons to fall into the potential wells created by the CDM density contrast. This causes their density contrast to grow until it matches the CDM density contrast. On smaller scales that does not happen, and instead the acoustic oscillation continues though damped by the diffusion of the baryon.

For wavelengths that enter the horizon after last scattering, the behaviour is much simpler. All that happens is the increase in the matter density contrast.

13.2 BARYON ACOUSTIC OSCILLATION

In the rest of this chapter I deal with the baryon acoustic oscillation. In a useful approximation, we can take photon diffusion to be completely absent so that the number of baryons per photon n_b/n_γ is constant at each location. This is called the tight-coupling approximation.

According to the tight coupling approximation, the perturbed n_b/n_γ is a constant, just like the unperturbed quantity. It follows that the perturbation $\delta(n_b/n_\gamma)$ is constant, which leads to a useful relation between the baryon and photon density contrasts. To derive it, we can invoke the local scale factor $a(\mathbf{x}, t)$. In terms of it we have

$$n_\gamma \propto n_b \propto \rho_b \propto a^{-3}, \qquad \rho_\gamma \propto a^{-4}. \tag{13.1}$$

This gives $n_b \propto \rho_b$ but $n_\gamma \propto \rho_\gamma^{3/4}$. It follows that

$$\frac{n_\gamma}{n_b}\delta(n_b/n_\gamma) = \frac{\delta n_b}{n_b} - \frac{\delta n_\gamma}{n_\gamma} \tag{13.2}$$

$$= \delta_b - \frac{3}{4}\delta_\gamma \tag{13.3}$$

is constant. According to Eq. (11.33), the constant is zero with the adiabatic condition that we are adopting. We conclude that the tight-coupling approximation gives

$$\delta_b = \frac{3}{4}\delta_\gamma. \tag{13.4}$$

Also, $V_b = V_\gamma$ and the pressure comes almost entirely from the photons. We can therefore deal just with δ_γ and V_γ.

Since the typical photon energy is much less than the electron's rest energy, its scattering off an electron leaves its energy almost unchanged (Thomson scattering as opposed to Compton scattering). The photon energy therefore satisfies the energy continuity equation which gives[2]

$$\dot{\delta}_\gamma = -\frac{4}{3}kV_\gamma + 4\dot{\Phi}. \tag{13.5}$$

(As in the previous chapter, an overdot in this chapter denotes $d/d\eta = ad/dt$.) The acceleration equation is

$$\dot{V}_\gamma = -aH(1-3w_f)V_\gamma - \frac{\dot{w}_f}{1+w_f}V_\gamma + \frac{k}{3}\frac{\rho_\gamma}{\rho_b + \frac{4}{3}\rho_\gamma}\delta_\gamma + k\Psi. \tag{13.6}$$

where w_f (referring to the baryon-photon fluid) is equal to $\frac{1}{3}\rho_\gamma/(\rho_\gamma+\rho_B)$. Combining these two equations gives

$$\frac{1}{4}\ddot{\delta}_\gamma + \frac{1}{4}\frac{R}{1+R}\dot{\delta}_\gamma + \frac{1}{4}k^2c_s^2\delta_\gamma = F, \tag{13.7}$$

where

$$F = -\frac{k^2}{3}\Psi + \frac{\dot{R}}{1+R}\dot{\Phi} + \ddot{\Phi} \tag{13.8}$$

$$c_s^2 \equiv \frac{\dot{P}_f}{\dot{\rho}_f} = \frac{\dot{P}_\gamma}{\dot{\rho}_\gamma + \dot{\rho}_b} = \frac{1}{1+R}, \tag{13.9}$$

with

$$R \equiv \frac{3}{4}\frac{\rho_b}{\rho_\gamma} \tag{13.10}$$

$$= \frac{3}{4}\frac{\rho_b}{\rho_m}\frac{\rho_r}{\rho_\gamma}\frac{\rho_m}{\rho_r} \tag{13.11}$$

$$= \frac{3}{4}\frac{\rho_b}{\rho_m}\frac{\rho_\gamma + \rho_\nu}{\rho_\gamma}\frac{a(\eta)}{a_{eq}} \tag{13.12}$$

$$= 0.034\frac{a(\eta)}{a_{eq}}. \tag{13.13}$$

(To obtain the final expression I used the numbers given in Table 6.1.)

If F were zero we would be dealing with a damped oscillator equation. But Eqs. (12.10), (12.12), and (12.13) give F as a linear combination of δ_γ, $\dot{\delta}_\gamma$, $\ddot{\delta}_\gamma$ and some slowly varying terms. As a result, according to these equations, δ_γ undergoes a damped oscillation, which is pushed off-centre by the slowly varying terms.

As Eqs. (13.7) and (13.8) involve only k as opposed to $\mathbf{k}$, the initial condition[3] Eq. (12.21) will lead to a result of the form

$$\delta_\gamma(\mathbf{k},\eta)e^{i\mathbf{k}\cdot\mathbf{x}} + \delta_\gamma(-\mathbf{k},\eta)e^{-i\mathbf{k}\cdot\mathbf{x}} = y(k,\eta)\left[\zeta(\mathbf{k})e^{i\mathbf{k}\cdot\mathbf{x}} + \zeta(-\mathbf{k})e^{-i\mathbf{k}\cdot\mathbf{x}}\right], \tag{13.14}$$

---

[2]The last term of the corresponding equation (7.1) is absent because $P_\gamma = \rho_\gamma/3$.
[3]Since the initial epoch is radiation dominated, we then have $\delta = \delta_\gamma$.

with y depending only on k. The acoustic oscillation is therefore a standing wave. Each Fourier component is of the form

$$\frac{1}{4}\delta_\gamma(\mathbf{k}, \eta) = A(\mathbf{k}, \eta) + B(\mathbf{k}, \eta)\cos(kr_\mathrm{s}(\eta)) + C(\mathbf{k}, \eta)\sin(kr_\mathrm{s}(\eta)), \qquad (13.15)$$

with A, B and C slowly varying and

$$r_\mathrm{s}(\eta) = \int_0^\eta c_\mathrm{s}(\eta)d\eta. \qquad (13.16)$$

To estimate A we should drop in Eqs. (13.7) and (13.8) all terms that are differentiated with respect to η, giving

$$A(\mathbf{k}, \eta) = -(1 + R(\eta))\Phi(\mathbf{k}). \qquad (13.17)$$

Although Eqs. (13.7) and (13.8) give some damping, the dominant damping comes from an effect that we have so far ignored. This is the effect of photon diffusion, which is called Silk damping after Joseph Silk who first pointed to its existence. To calculate the effect of Silk damping, we need the cross section for Thomson scattering:

$$\sigma_\mathrm{T} = \frac{8\pi}{3}\left(\frac{\alpha}{m_\mathrm{e}}\right)^2 = 6.65 \times 10^{-25}\,\mathrm{cm}^2. \qquad (13.18)$$

As a result of Thomson scattering, each photon performs a random walk. The mean average time for a step is $t_\mathrm{s} \sim (cn_\mathrm{e}\sigma_\mathrm{T})^{-1}$ where n_e is the electron number density.[4] The mean number of steps in time t is $N = t/t_\mathrm{s}$, and in that time a photon diffuses a distance $\sqrt{N}ct_\mathrm{s} \sim c(tt_\mathrm{s})^{1/2}$. At a given epoch, Silk damping therefore reduces, and effectively removes, the oscillation for a/k bigger than

$$\frac{a}{k_\mathrm{s}} \simeq c(tt_\mathrm{s})^{1/2} \simeq \left(\frac{ct}{cn_\mathrm{e}\sigma_\mathrm{T}}\right)^{1/2}. \qquad (13.19)$$

The acoustic oscillation ceases at the epoch of last scattering. The relevant parameters are then $R = 0.78$, $r_\mathrm{s} = 145\,\mathrm{Mpc}$ and $a/k_\mathrm{s} = 8\,\mathrm{Mpc}$. The spectrum $\mathcal{P}_{\delta_\gamma}(k)$ of the photon density contrast at last scattering exhibits a series of peaks and troughs, caused by the acoustic oscillation. For the wavelength corresponding to the first trough, there has been time for one quarter of an oscillation; for the wavelength corresponding to the first peak there has been time for half an oscillation, and so on. The spectrum

$$\mathcal{P}_{\delta_\mathrm{b}}(k) = (9/16)\mathcal{P}_{\delta_\gamma}(k) \qquad (13.20)$$

of the baryon density contrast exhibits the same features.

---

[4]The initial and final states are occupied in our case which in principle affects the transition probability, but it can be shown that the effect is small.

EXERCISES

1. Verify Eqs. (13.7) and (13.17).

2. When the neutrinos are ignored, the tight-coupling approximation provides a complete system of equations that can be solved if $\zeta(\mathbf{k})$ is known. Write down these equations, and the initial conditions that should be invoked to solve them.

3. Verify the factor 0.034 in Eq. (13.13).

4. How slow is the variation of A, B and C, which justifies Eq. (13.14)?

5. Explain the origin of the factor 9/16 in Eq. (13.20).

6. By considering a tube with cross section area σ_T, verify the formula for t_s given after Eq. (13.18).

CMB anisotropy

CONTENTS

Now we come to the CMB anisotropy. We are interested in the intrinsic anisotropy, defined in the rest frame of the CMB, which is of two kinds. First, there is the anisotropy in its intensity shown in Figure 14.1. Second, there is polarization. As with the spatial perturbations, we are interested in the statistical properties of the CMB anisotropy.

In dealing with the CMB anisotropy I will discount the effect of galaxy clusters. That effect is easily allowed for in the case of clusters that are near enough to identify individually, and calculation shows that the effect of more distant clusters is negligible except on very small scales where the rest of the anisotropy becomes negligible.

14.1   CMB TEMPERATURE

Before the epoch of last scattering, corresponding to the formation of hydrogen atoms, the photons are in thermal equilibrium and satisfy the blackbody distribution. If the formation were instantaneous the photons would retain precisely the blackbody distribution. In reality it is gradual, but because there are so many photons per baryon. the blackbody distribution is still retained to very high accuracy. That is the case for the homogeneous Universe, and it remains the case when perturbations are taken into account. The perturbations, though, mean that the temperature of the blackbody distribution depends on the direction of observation. We therefore deal with a temperature $T + \Delta T$, where the unperturbed quantity T may be taken to be the average over all directions.

The perturbation ΔT comes from three sources: (i) the inhomogeneity in the photon temperature at the epoch of last scattering, (ii) the inhomogeneity in the motion of the photon fluid at last scattering which causes the redshift of the CMB to depend on direction, and (iii) the effect of perturbations in the matter density along the line of sight which is called the Sachs-Wolfe effect. The first two obviously

Intrinsic anisotropy of the CMB

Figure 14.1 The image is a map of the whole sky. The boundary of the map corresponds to a single direction in the sky, and its centre corresponds to the opposite direction. The variation in brightness is only about one part in 10, 000.

depend only on conditions at the surface of last scattering. It can be shown that this is true to a good approximation also for the Sachs-Wolfe effect, which in any case affects ΔT only on large angular scales.[1]

After last scattering, the photons in the standard cosmology travel freely which means that their blackbody distribution is retained. A departure from the blackbody distribution would indicate a departure from the standard cosmology, such as the presence of a decaying particle species. The fact that no departure from the blackbody distribution has been observed constrains possible departures from the standard cosmology. For instance, if a decaying species is supposed to have some definite decay rate, it constrains the number density of the decaying particles.

14.2 CMB MULTIPOLES

Within the present accuracy of observation, the CMB anisotropy can be described using linear perturbation theory. According to linear perturbation theory, the anisotropy in the intensity of the CMB corresponds to an anisotropy in its temperature. It is convenient to work with the multipoles, defined by

$$\Delta T(\mathbf{e}) = \sum_{\ell m} a_{\ell m} Y_{\ell m}(\theta, \phi), \qquad (14.1)$$

---

[1]To be precise the Sachs-Wolfe effect dominates when the ℓ that we consider shortly is $\lesssim 30$ but becomes negligible for much bigger ℓ.

where $\mathbf{e}$ is the direction in the sky corresponding to spherical polar angles $\{\theta, \phi\}$. The monopole $\ell = 0$ is discounted because it can be regarded as belonging to the isotropic part of the CMB, and the dipole is discounted because it just corresponds to our motion with respect to the CMB rest frame. That motion corresponds to a speed $v \sim 10^{-3}c$, and to first order in v/c the motion has no effect on the $\ell \geq 2$ multipoles. Within the accuracy of present observations, there is no need to go to a higher order in v/c. In other words, we can take the observed $\ell \geq 2$ multipoles to refer to the rest frame of the CMB.

As we saw in the previous section, $\Delta T(\mathbf{e})$ depends mostly on conditions at the surface of last scattering. From the properties of the $Y_{\ell m}$, it follows that a multipole with given $\ell \gg 1$ comes mostly from cosmological perturbations with wavenumber k given by[2]

$$\ell \sim x_{\text{ls}} k. \tag{14.2}$$

where x_{ls} is the comoving distance to the last-scattering surface, practically the same as the distance to the horizon which is $14.0\,\text{Gpc}$.

As with the spatial perturbations, the statistical properties of the CMB are fundamentally defined with respect to some ensemble of universes, but statistical homogeneity means that the ensemble average can also be regarded as a position average. In other words, we can define the statistical properties of the multipoles by averaging over the position of the observer of the CMB. The average of a quantity is denoted by $\langle \rangle$.

The statistical properties of the CMB anisotropy are supposed to be invariant under rotations (statistical isotropy). This requires $\langle a_{\ell m} \rangle = 0$. To see what it requires for $\langle a_{\ell m} a^*_{\ell m} \rangle$ we have to consider the behaviour of $Y_{\ell,m}(\theta, \phi)$ under rotations. Recall that the spherical polar angles are defined in terms of Cartesian coordinates by

$$x = r \sin\theta \cos\phi \tag{14.3}$$
$$y = r \sin\theta \sin\phi \tag{14.4}$$
$$z = r \cos\theta, \tag{14.5}$$

where the choice of r is arbitrary. If the Cartesian coordinates are rotated to give new angles $\{\theta', \phi'\}$, then $Y_{\ell,m}(\theta, \phi)$ in a given direction undergoes a unitary transformation

$$Y_{\ell m}(\theta', \phi') = \sum_{m'} U_{mm'}(\ell) Y_{\ell m'}(\theta, \phi). \tag{14.6}$$

As $\delta T(\mathbf{e})$ doesn't change, the multipoles undergo the inverse transformation

$$a'_{\ell m} = \sum_{m'} U_{m'm}(\ell) a_{\ell m'}. \tag{14.7}$$

---

[2]This relation comes up also in quantum mechanics. The wave-function of a particle with angular momentum ℓ has angular dependence $Y_{\ell m}$, and for $\ell \gg 1$ we can construct a wave packet built from these functions with almost definite ℓ, which is at some distance x and is moving in a transverse direction and has momentum k. Then ℓ is given to good accuracy by the classical relation $\ell = xk$.

Invariance under rotations therefore requires $\langle a_{\ell m} a_{\ell' m'}^* \rangle$ to be of the form

$$\langle a_{\ell m} a_{\ell' m'}^* \rangle = \delta_{\ell \ell'} \delta_{mm'} C_\ell. \tag{14.8}$$

The quantity C_ℓ is called the spectrum of the CMB anisotropy.

14.3 POLARIZATION

Before the epoch of last scattering, the photons are frequently scattering off electrons, and after that epoch they travel freely. If the transition from one situation to the other were instantaneous, the CMB would be unpolarized. In reality the transition take some time, and during that time the Thomson scattering of photons off electrons generates polarization. The polarization is of interest because it may offer the best chance of detecting the effect of the gravitational radiation created during inflation.

To deal with the polarization of the CMB, we need the polarization tensor. To define it we consider the CMB coming from an infinitesimal patch of sky around some direction $\mathbf{e}$, and from an infinitesimal range of angular frequencies around some angular frequency ω. The corresponding electric field is orthogonal to $\mathbf{e}$ and I write it as

$$\mathbf{E(t)} = \mathrm{Re}\left[\mathbf{E}e^{i\omega t}\right]. \tag{14.9}$$

To describe it I use coordinates $\{x^1, x^2\}$ where x^1 is in the θ direction of some spherical coordinate system and x^2 is in the ϕ direction. Then I write for the complex amplitude $\mathbf{E}$

$$E_i^* E_j + E_j^* E_i = 2\left(I\delta_{ij} + 4P_{ij}\right), \tag{14.10}$$

where E_1 and E_2 are the components of the complex amplitude $\mathbf{E}$ and P_{ij} is traceless. The intensity of the wave is I, and P_{ij} is called the polarization tensor. Following the usual convention when considering the CMB, a factor 4 is inserted in front of P, which is motivated by the fact that as we change the direction of observation, $\delta I/I = 4\delta T/T$. Using linear perturbation theory, one can show that P_{ij} is independent of ω, just like $\delta I/I$.

Since P_{ij} is traceless and symmetric we can write it in the form

$$P = \begin{pmatrix} Q & U \\ U & -Q \end{pmatrix}, \tag{14.11}$$

and it is convenient to define $Q_\pm = Q \pm iU$. Then, as seen in Appendix A.1, one can write

$$Q_\pm(\mathbf{e}) = \sum_{\ell=2}^{\infty} \sum_{m=-\ell}^{\ell} Q_{\ell m}^\pm \, {}_{\pm}Y_{\ell m}, \tag{14.12}$$

where the functions ${}_{\pm}Y_{\ell m}$ take care of the behaviour of $Q_\pm$ under rotations of the coordinate system.

Finally, we write $Q_{\ell m}^{\pm} = E_{\ell m} \pm iB_{\ell m}$. One can show that $B_{\ell m}$ could be caused only by gravitational radiation, generated during inflation as described in Chapter 19. The fact that it has not yet been detected places an upper bound on the energy density during inflation. In contrast, $E_{\ell m}$ has been observed. Analogously with Eq. (14.8) we can write

$$\langle E_{\ell m}^* E_{\ell' m'} \rangle = C_\ell^{EE} \delta_{\ell\ell'} \delta_{mm'} \tag{14.13}$$

$$\langle a_{\ell m}^* E_{\ell' m'} \rangle = C_\ell^{TE} \delta_{\ell\ell'} \delta_{mm'}. \tag{14.14}$$

14.4   WHAT THE SPECTRA TELL US

The measured values of the spectra C_ℓ, C_ℓ^{EE} and C_ℓ^{TE} are shown in Figures 14.2–14.4. The troughs and peaks are caused by the baryon acoustic oscillation present at last scattering. Also shown are lines corresponding to a best fit to the measured values, obtained from the Boltzmann equations by varying the seven fundamental cosmological parameters listed in Table A.3. The values and uncertainties of the parameters that are shown in Table A.3 are the outcome of this fit. Very remarkably, they are not significantly altered if we take into account all other observations.

The spectra of the CMB anisotropy, all by themselves, are therefore superseding decades of work that has gone into the determination of the parameters by other methods. Take for instance the Hubble constant H_0. It was traditionally determined using Hubble's law, but to do that one has to determine the distances of galaxies as shown in Figure 6.1. To make that determination one has to construct what is called the cosmic distance ladder. The cosmic distance ladder begins with the determination of the distance of nearby objects, by observing their parallax (the change in their direction as the Earth moves in its orbit around the sun). Among these objects, one identifies a type of object which has always the same luminosity, called a standard candle. Then, by observing the brightness of the standard candle in other galaxies one deduces their distance, even though it is too big to generate observable parallax. The process can be repeated; a second and brighter standard candle can by identified and used to determine the distance of galaxies that are too far away to permit the use of the original standard candle, and so on. This technique for measuring distance is called the distance ladder. As one may guess from its construction the distance ladder is quite shaky, and the determination of H_0 using it has a bigger uncertainty than the one coming from the CMB anisotropy.

There is a concern that one might have in using the spectra of the CMB anisotropy, but it turns out not to matter. The concern is that, say, $C_\ell \equiv \langle |a_{\ell m}|^2 \rangle$ is an average over the position of the observer, while we measure only a single $a_{\ell m}$. The mean square difference between the two is

$$(\Delta C_\ell)^2 \equiv \langle (|a_{\ell m}|^2 - C_\ell)^2 \rangle = \langle |a_{\ell m}|^4 - C_\ell^2 \rangle. \tag{14.15}$$

This is called the cosmic variance of $|a_{\ell m}|^2$. To evaluate it, we need to remember

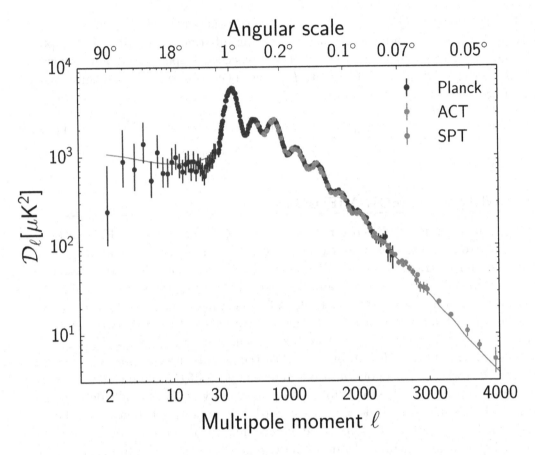

Figure 14.2 The quantity plotted is $\ell(\ell+1)C_\ell/2\pi$. For $\ell > 30$, the quantity has been binned.

that, as seen in Chapter 11, the Fourier components of perturbations are Gaussian to high accuracy. Within the linear approximation, the multipoles will inherit the Gaussianity property of the spatial perturbations. As a result, their real and imaginary parts will have Gaussian probability distributions. If a quantity g has a Gaussian probability distribution we have

$$\langle g^4 - \langle g^2 \rangle^2 \rangle = 2\langle g^2 \rangle^2. \tag{14.16}$$

If only a single real or imaginary part of an $a_{\ell m}$ were measured, the cosmic variance would be $2C_\ell^2$. But with all $(2\ell+1)$ of the $a_{\ell m}$ measured we have $(\Delta C_\ell)^2 = C_\ell^2/(2\ell+1)$. Cosmic variance is therefore significant only for very small ℓ, which means that it has a negligible effect on the determination of the cosmological parameters.

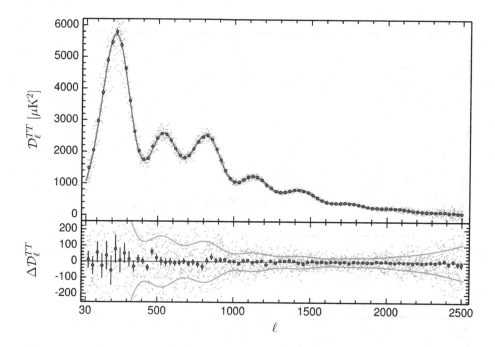

Figure 14.3 The error bars are the same as in Figure 14.2, but showing only $\ell > 30$. The measured values of the individual C_ℓ's are now shown as dots. The lines in the lower figure are 68% confidence level uncertainties for the individual C_ℓ's.

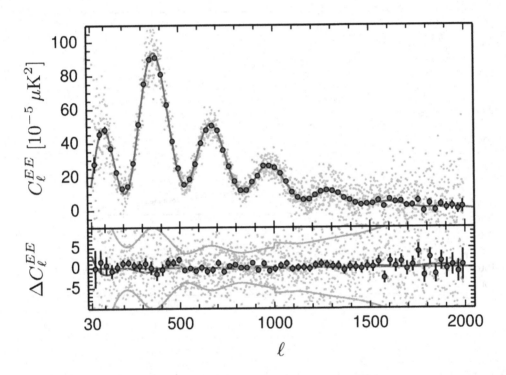

Figure 14.4 This is the same as Figure 14.3 except that it shows C_ℓ^{EE}.

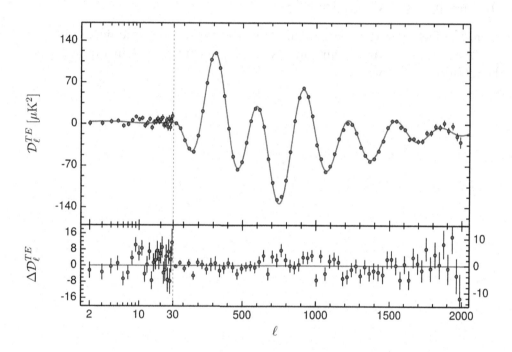

Figure 14.5 The quantity plotted is $\ell(\ell + 1)C_\ell^{\mathrm{TE}}/2\pi$. For $\ell > 30$, the quantity has been binned.

EXERCISES

1. Use Eq. (A.5) to show that rotational invariance requires Eq. (14.8).

2. Verify the second equality in Eq. (14.15).

3. Define the Gaussian probability distribution. Write down the definite integral which gives the normalization of this distribution, and the definite integral which gives Eq. (14.16).

4. Estimate the spacing of peaks of $\mathcal{P}_\gamma(k)$, evaluate at last scattering as described at the end of Chapter 13. Use Eq. (14.2) to show that it is roughly consistent with the spacing of the peaks of C_ℓ.

5. Using the fact that the energy density of the electromagnetic field is $(E^2 + B^2)/2$ at any spacetime point, justify the statement that I in Eq. (12.3) is the intensity of the wave.

Galaxy formation

CONTENTS

In this chapter I describe galaxy formation. As already mentioned in Chapter 5, a galaxy forms when some slightly over-dense region of the Universe attracts more matter to collapse under its own weight.

For both the CDM and the baryons, the energy density is practically equal to c^2 times the mass density and it makes more sense to deal with the latter. The density contrast δ, previously defined as the energy density perturbation divided by the average energy density, can as well be defined as the mass density perturbation divided by the average mass density.

15.1   MATTER TRANSFER FUNCTION

In this section I define an object $T(k)$, called the matter transfer function. Then I obtain an approximate expression for it, and display the exact result.

To arrive at $T(k)$ we need to consider the behaviour of the matter density contrast during matter domination. During matter domination Eq. (12.13) gives $\Phi = \Psi$. The CDM doesn't interact and has negligible pressure, and so analogously with Eqs. (12.10) and (12.11) it satisfies

$$\dot{\delta}_{\mathrm{c}} = -kV_{\mathrm{c}} + 3\dot{\Phi} \tag{15.1}$$

$$\dot{V}_{\mathrm{c}} = -aHV_{\mathrm{c}} + k\Psi. \tag{15.2}$$

The total matter density contrast δ also satisfies Eqs. (15.1) and (15.2), for wavelengths bigger than the Jeans wavelength. In addition, δ related to Φ by Eq. (12.15). The growing solution of Eqs. (12.15), (15.1), and (15.2) is then $\delta \propto a$. From Eq. (12.15) this corresponds to constant Φ and

$$\delta(\mathbf{k}, t) = -\frac{2}{3}\left(\frac{k}{aH}\right)^2 \Phi(\mathbf{k}). \tag{15.3}$$

This is the same as Eq. (10.28) which was derived in the Newtonian regime $k \gg aH$. We have now extended its range of validity to include also $k \lesssim aH$.

For scales $1/k$ corresponding to wavelengths that are far outside the horizon at the onset of matter domination, Eq. (12.20) applies with good accuracy, giving $\Phi(\mathbf{k}) = -(3/5)\zeta(\mathbf{k})$. The transfer function $T(k)$ is defined by

$$\Phi(\mathbf{k}) = -\frac{3}{5}T(k)\zeta(\mathbf{k}). \tag{15.4}$$

To estimate $T(k)$ we need to consider wavelengths that enter the horizon during radiation domination. At this epoch we will have roughly

$$\dot{\delta}_c \simeq aH\delta_c, \tag{15.5}$$

because H^{-1} is the only relevant distance scale.

Since δ_ν decays, $\delta \simeq \delta_\gamma$ for as long as there is radiation domination. As we saw in Chapter 13, δ_γ oscillates with decreasing amplitude, which from Eq. (12.15) implies that Φ decreases rapidly. It soon becomes negligible and then Eqs. (15.1) and (15.2) give

$$\ddot{\delta}_c \simeq -aH\dot{\delta}_c. \tag{15.6}$$

With the initial conditions (12.21) and (15.5) the solution of this equation is $\delta_c \simeq \zeta \ln(k/aH)$. This holds until around the time of matter-radiation equality. Denoting that epoch by a subscript eq, we have

$$\delta_c \simeq \zeta \ln(k/k_{\mathrm{eq}}), \tag{15.7}$$

where $k_{\mathrm{eq}} = (aH)_{\mathrm{eq}}$. At this time, $|\delta_b| \ll |\delta_c|$ because δ_b has been undergoing the acoustic oscillation while $|\delta_c|$ has been increasing. Setting $\delta_b = 0$ we have $\delta = (\rho_c/\rho_m)\delta_c$ and can set $\delta = \delta_c$ for an approximate calculation. Then Eqs. (12.15) and (15.7) give

$$T(k) \simeq \frac{k_{\mathrm{eq}}^2}{k^2} \ln\left(\frac{k}{k_{\mathrm{eq}}}\right). \qquad (k \gtrsim k_{\mathrm{eq}}). \tag{15.8}$$

The exact transfer function, calculated using the Boltzmann equation, is shown in Figure 15.1. The approximation (15.8) is roughly correct, but looking closely one can see an irregularity which is caused by the acoustic oscillation.

15.2   FORMATION OF CDM HALOS

As we saw in Chapter 5, the CDM outweighs the baryons. As a result, the CDM density contrast during matter domination evolves almost as if there were no baryons. To discuss that evolution, I will work with the smoothed density contrast defined in Section 11.3.

The smoothed density contrast grows with time because over-dense regions attract more matter towards them. The growth can begin only when the smoothing radius enters the horizon. During matter domination, the CDM density contrast is proportional to the scale factor a. Using Eqs. (10.21) and (11.24), its Fourier components during matter domination on scales bigger than the Jeans scale are given by

$$\delta(\mathbf{k}, z, R) = \frac{2}{5} W(kR) \left(\frac{k}{H_0}\right)^2 \frac{T(k)}{\Omega_\mathrm{m}} \frac{1}{1+z} \zeta(\mathbf{k}). \qquad (15.9)$$

The mean square smoothed density contrast is given by

$$\sigma^2(z, R) = \left(\frac{1}{1+z}\right)^2 \sigma_0^2(R) \qquad (15.10)$$

$$\sigma_0^2(R) \equiv \int_0^\infty \frac{dk}{k} \left[\frac{2}{5} W(kR) \left(\frac{k}{H_0}\right)^2 \frac{T(k)}{\Omega_\mathrm{m}}\right]^2 \mathcal{P}_\zeta(k). \qquad (15.11)$$

The smoothed density contrast has over-dense and under-dense regions. It is important for the development of structure that each such region is roughly as big as the smoothing radius. That is the case because the smoothed density contrast includes only Fourier components whose wavelength is bigger than the smoothing radius. Of these, the dominant ones have wavelength not *much* bigger than the smoothing radius, because they have spent the most time with wavelength less than the Hubble distance, allowing them to grow the most.

When the CDM's density contrast becomes roughly equal to 1 in some region, the CDM in that region collapses under its own weight. As the collapse proceeds, the inward motion of the CDM particles is randomised by the effect of gravity, which halts the collapse when the collapsing region is about half as big as it was when the collapse started. We then have an object of fixed size, which is called a CDM halo.

To estimate the mass M of the CDM halo, we can pretend that the collapsing region is a sphere. Well before the collapse starts, the sphere is expanding smoothly, with radius aR where R is the smoothing scale. The mass of matter within the region is

$$M(R) = \frac{4}{3}\pi R^3 a^3 \rho_\mathrm{mat} = \frac{4}{3}\pi R^3 \rho_{0\mathrm{mat}} \qquad (15.12)$$

$$= 5.37 \times 10^{11} \left(\frac{R}{\mathrm{Mpc}}\right)^3 M_\odot, \qquad (15.13)$$

where ρ is the almost homogeneous energy density at early times and the subscript 0 denotes the present.

At first, the collapse occurs only at places that represent exceptionally high peaks of the density contrast. That doesn't significantly affect the density contrast at other locations, which continues to be proportional to the scale factor a. Eventually though, the collapse occurs even at peaks that are not exceptionally high. A large fraction of the matter in the universe then collapses and we can no longer talk about a perturbation in the density of the CDM. On a given scale, the final collapse occurs when $\sigma(z, R) \simeq 1$, which occurs at the redshift given by $(1 + z) \simeq \sigma_0(R)$. Figure 15.2 shows $\sigma_0(R)$ as a function of the mass $M(R)$. We see that the CDM halo formation begins at $z \sim 10$, with the formation of the lightest halos. After that, successively heavier CDM halos form. The heaviest CDM halos, with mass $\sim 10^{14} M_\odot$, form at $z \simeq 1$. Soon after that, the cosmological constant starts to accelerate the expansion which prevents the formation of CDM halos.

15.3 FORMATION OF GALAXIES AND CLUSTERS

So much for the CDM. What about the baryons? At the epoch of last scattering their smoothed density contrast is much smaller than that of the CDM, because it has been undergoing the acoustic oscillation instead of growing. After last scattering, gravity tries to pull the baryons into regions where the CDM is over-dense. On the other hand, the random motion of the baryons tries to prevent that from happening. As we have seen in Chapter 10, gravity wins if the mass of the final collapsed object is bigger than about $6 \times 10^3 M_\odot$. That is our estimate for the mass of the lightest galaxies, which is in accordance with observation.

If gravity does not win we have a pure CDM halo, but no such object has been observed. That is probably because pure CDM halos collide with each other, to form halos which are big enough to become galaxies.

As we saw, the collapse of a CDM halo is halted when the random motion of the CDM particles becomes big enough to counteract the effect of gravity. The same is true for the collapse of the baryons, but the baryons collapse further because they emit electric radiation which tends to reduce the random motion. As a result, the baryons in a galaxy are confined to a luminous central region, with the CDM halo around it. That's why we call it a CDM *halo*.

The smallest galaxies form first. Successively bigger galaxies form through the collapse of successively bigger CDM halos, and also through the merging of existing galaxies. This process of direct galaxy formation ends with the formation of the biggest galaxies. After that, the would-be galaxy is too big to survive and the collapsing region breaks up into galaxy-sized pieces so that we have a galaxy cluster. The mass of the cluster is given by Eq. (15.13). Successively larger galaxy clusters form in this way, until we arrive at the epoch when the cosmological constant starts to become significant. The process of gravitational collapse then ceases, because the gravitational effect of the cosmological constant is repulsive rather than attractive, as we saw in Chapter 7. If the cosmological constant lives up to its name, and indeed remains constant in the future, there will never be any gravitationally bound objects bigger than the biggest galaxy clusters.

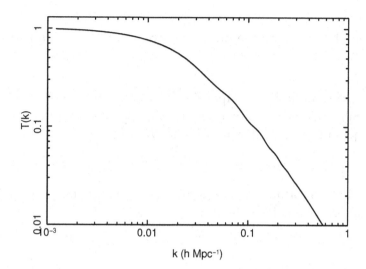

Figure 15.1 The matter transfer function. The slight irregularity is caused by the baryon acoustic oscillation. Reproduced from *'The Primordial Density Perturbation'* by DHL and A. R. Liddle.

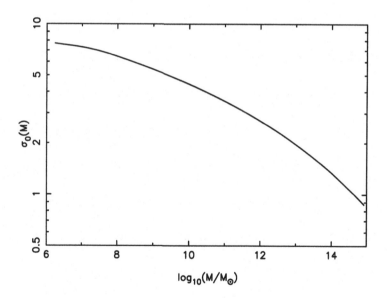

Figure 15.2 The quantity $\sigma_0(R(M))$ defined in the text. The formation of CDM halos with mass M occurs at redshift given by $1 + z \simeq 1/\sigma_0(R(M))$. Reproduced from 'The Primordial Density Perturbation' by DHL and A. R. Liddle.

EXERCISES

1. Using Table 11.1, estimate the range of k over which $T(k)$ will be close to 1.

2. Verify Eq. (15.9).

3. Check that the result $a/k_s = 8\,\text{Mpc}$, given in the text for the inverse wavenumber of the Silk scale at last scattering, is roughly consistent with the spacing of the bumps in the curve shown in Figure 15.1.

4. Verify Eq. (15.13).

5. Use Figure 15.2 to estimate the mass of an object forming at redshift $z = 4$. Is that a galaxy, or a galaxy cluster?

III

Inflation

Scalar fields: Classical theory

CONTENTS

The electric and magnetic fields are vector fields, which means that one has to specify both their strengths and their directions. There also exist scalar fields, specified entirely by their strength. One of them is the Higgs field, whose oscillation corresponds to the Higgs boson. There is, almost certainly, at least one more scalar field. That is the field responsible for inflation, called the inflaton field. In this chapter I describe scalar fields at the classical level, and in the next I include quantum physics. I use natural units.

16.1 ACTION PRINCIPLE

As far as we know, the fundamental laws of physics can be derived from an **action** S. In flat spacetime it has the form

$$S = \int_{-\infty}^{\infty} L \, dt \, . \tag{16.1}$$

The Lagrangian L depends on the quantities that specify the state of the system under consideration. These quantities are called degrees of freedom.

The classical evolution of the degrees of freedom is given by the action principle. This states that $\delta S = 0$, where δS is the change in S resulting from a small change in the time dependence of the degrees of freedom. In this section, we see how the action principle works without attaching a physical meaning to the degrees of freedom.

We begin by assuming a single degree of freedom, denoted by q. The Lagrangian is taken to be a function of q and its time derivative $\dot{q}$, and for the moment we allow it to also depend explicitly on time. Then we have

$$\delta S = \int_{-\infty}^{\infty} dt \left(\frac{\partial L}{\partial q} \delta q + \frac{\partial L}{\partial \dot{q}} \delta \dot{q} \right) . \tag{16.2}$$

Integrating the second term by parts we have

$$\delta S = \int_{-\infty}^{\infty} dt \left[\frac{\partial L}{\partial q} - \frac{\partial}{\partial t} \left(\frac{\partial L}{\partial \dot{q}} \right) \right] \delta q + \int_{-\infty}^{\infty} dt \frac{\partial}{\partial t} \left(\frac{\partial L}{\partial \dot{q}} \delta q \right). \tag{16.3}$$

The second term of this expression vanishes if δq vanishes sufficiently rapidly at infinity, which we demand. Then the action principle gives the Lagrangian equation of motion

$$\frac{\partial L}{\partial q} - \frac{\partial}{\partial t} \left(\frac{\partial L}{\partial \dot{q}} \right) = 0. \tag{16.4}$$

The Hamiltonian is defined as

$$H(q, p, t) \equiv p\dot{q} - L, \tag{16.5}$$

with $p \equiv \partial L / \partial \dot{q}$. From Eq. (16.4) one can derive the Hamiltonian equations of motion:

$$\dot{q} = \frac{\partial H}{\partial p} \qquad \dot{p} = -\frac{\partial H}{\partial q}. \tag{16.6}$$

Using the chain rule, these give in turn the time dependence of any quantity $A(q, p, t)$:

$$\frac{dA}{dt} = \sum_n \left(\frac{\partial H}{\partial p} \frac{\partial A}{\partial q} - \frac{\partial A}{\partial p} \frac{\partial H}{\partial q} + \frac{\partial A}{\partial t} \right). \tag{16.7}$$

Applying this equation to H, we see that H is time independent (conserved) if L has no explicit time dependence. It is then, by definition, the energy of the system. The absence of explicit time dependence means that the theory is invariant under time displacements, $q_n(t) \to q_n(t + T)$ where T is any constant. Therefore *energy is conserved if and only if the theory is invariant under time displacements*. From now on I assume that this is the case, so that L has no explicit time dependence.

In the context of field theory, one is particularly interested in a Lagrangian of the form

$$L = \frac{1}{2}\dot{q}^2 - \frac{1}{2}\omega^2 q^2. \tag{16.8}$$

In this case, $p = \dot{q}$ and we have

$$H = \frac{1}{2}p^2 + \frac{1}{2}\omega^2 q^2, \tag{16.9}$$

and each degree of freedom satisfies the harmonic oscillator equation

$$\ddot{q}(t) = -\omega^2 q(t). \tag{16.10}$$

Repeating the discussion for more degrees of freedom denoted $q_1, q_2, \cdots$, we have

$$\frac{\partial L}{\partial q_n} - \frac{\partial}{\partial t} \left(\frac{\partial L}{\partial \dot{q}_n} \right) = 0, \tag{16.11}$$

and

$$H(q_n, p_n) \equiv \sum p_n \dot{q}_n - L, \qquad (16.12)$$

with $p_n \equiv \partial L / \partial \dot{q}_n$, and

$$\dot{q}_n = \frac{\partial H}{\partial p_n} \qquad \dot{p}_n = -\frac{\partial H}{\partial q_n}. \qquad (16.13)$$

In the case of harmonic oscillators we have

$$L = \sum_n \frac{1}{2}\dot{q}_n^2 - \frac{1}{2}\omega_n^2 q_n^2, \qquad (16.14)$$

giving $p_n = \dot{q}_n$ and

$$H = \sum_n \frac{1}{2}p_n^2 + \frac{1}{2}\omega_n^2 q_n^2, \qquad (16.15)$$

and each degree of freedom satisfies the harmonic oscillator equation

$$\ddot{q}_n(t) = -\omega^2 q_n(t). \qquad (16.16)$$

16.2 SCALAR FIELD

A scalar field ϕ depends on position $\mathbf{x}$. To invoke the action principle, one regards each $\phi(\mathbf{x})$ as a degree of freedom. The degrees of freedom are therefore labelled by a continuously varying quantity $\mathbf{x}$.

Let us begin with flat spacetime. To respect the relativity principle the action should be a scalar (Lorentz invariant). To achieve this, the Lagrangian for the fields must be of the form $L = \int \mathcal{L} \, d^3x$, where the **Lagrangian density** $\mathcal{L}$ is Lorentz invariant and has dimensions [energy]^4. The action is then

$$S = \int d^4x \, \mathcal{L} . \qquad (16.17)$$

We demand invariance under space-time translations, which means that $\mathcal{L}$ does not depend explicitly on the coordinates. Also, we demand that $\mathcal{L}$ depend only on ϕ and $\partial_\mu \phi$, so that

$$\delta S = \int d^4x \left[\frac{\partial \mathcal{L}}{\partial \phi} \delta\phi + \frac{\partial \mathcal{L}}{\partial(\partial_\mu \phi)} \delta(\partial_\mu \phi) \right]. \qquad (16.18)$$

Integrating the second term by parts, in each of the four variables x^μ, gives

$$\delta S = \int d^4x \left\{ \frac{\partial \mathcal{L}}{\partial \phi} - \frac{\partial}{\partial x^\mu} \left[\frac{\partial \mathcal{L}}{\partial(\partial_\mu \phi)} \right] \right\} \delta\phi + \int d^4x \frac{\partial}{\partial x^\mu} \left[\frac{\partial \mathcal{L}}{\partial(\partial_\mu \phi)} \delta\phi \right]. \qquad (16.19)$$

The second term of this expression vanishes if $\delta\phi$ vanishes sufficiently rapidly when each coordinate goes to infinity. Demanding that, the action principle gives

$$\frac{\partial \mathcal{L}}{\partial \phi} - \frac{\partial}{\partial x^\mu} \left(\frac{\partial \mathcal{L}}{\partial(\partial_\mu \phi)} \right) = 0 . \qquad (16.20)$$

This is the field equation for ϕ.

In the simplest case, $\mathcal{L}$ of the form

$$\mathcal{L} = -\frac{1}{2}\eta^{\mu\nu}\partial^{\mu}\phi\partial_{\nu}\phi - \frac{1}{2}m^2\phi^2 \,. \tag{16.21}$$

The scalar field equation is then

$$-\Box\phi + m^2\phi = 0\,, \tag{16.22}$$

where $\Box \equiv \delta^{\mu\nu}\partial_{\mu}\partial_{\nu}$ is the d'Alembertian discussed in Section 2.3. Eq. (16.22) is called the **Klein–Gordon equation.**

The Klein-Gordon equation has a constant solution, $\phi = 0$. This is the value of ϕ in the vacuum, called the vacuum expectation value or vev. The word 'expectation' is inserted because with quantum effects included the vacuum expectation value represents only an average, around which there are fluctuations in both time and space.

A time-dependent homogeneous field satisfies

$$\ddot{\phi} + m^2\phi = 0, \tag{16.23}$$

so that the field oscillates with angular frequency m. Discounting any homogeneous field, we can define Fourier components of ϕ by Eq. (10.17). The Fourier components satisfy

$$\ddot{\phi}(\mathbf{k}) + \left(k^2 + m^2\right)\phi(\mathbf{k}) = 0, \tag{16.24}$$

which means that they have angular frequency $\sqrt{k^2 + m^2}$. As the Fourier components do not interact with each other (or with the homogeneous term), ϕ in this case is called a free field.

More generally, the Lagrangian density is taken to have the form[1]

$$\mathcal{L} = -\frac{1}{2}\eta^{\mu\nu}\partial_{\mu}\phi\partial_{\nu}\phi - V(\phi)\,. \tag{16.25}$$

In this expression, $V(\phi)$ is some function, called the scalar field **potential.** The field equation is now

$$-\Box\phi + V'(\phi) = 0 \tag{16.26}$$

where the prime as usual denotes differentiation with respect to the argument.

For a homogeneous field

$$\ddot{\phi} + V'(\phi) = 0\,. \tag{16.27}$$

Eq. (16.26) is the equation for a non-relativistic unit-mass particle moving in one dimension, with position ϕ. That is why the potential $V(\phi)$ is so called. It is quite useful to keep in mind this analogy with particle mechanics.

In the simplest case $V(\phi)$ has a single minimum, which corresponds to the vev of ϕ. In general V may have more than one local minimum. Then there is the

---

[1]This form leads to a second-order evolution equation and seems to have been chosen by nature.

possibility that the field can move from one local minimum to another, through a quantum effect called tunnelling. The word 'tunnelling' is used because it is also used in the case of particle mechanics, where a particle initially located at one local minimum of the potential can after some time tunnel (with some probability) to a different local minimum. If one of the minima corresponds to a smaller V than any of the others, tunneling will in general drive ϕ to that minimum and it will correspond to the vev of ϕ. It might happen though that two or more local minima have the same V. It might also happen that the time taken for significant tunnelling from some higher local minimum to a lower one is longer than the age of the Universe. In these cases there is no physical principle that decides which particular minimum corresponds to the vev. It is as it is, just like the position of a particular galaxy.

In some region near the vev one will typically have to a good approximation

$$V(\phi) \simeq \frac{1}{2}m^2(\phi - \phi_0)^2, \tag{16.28}$$

with terms of cubic and higher order negligible. With the change of variable $\phi - \phi_0 \to \phi$, we recover the free field case.

The Hamiltonian is given by the generalization of Eq. (16.5), to the case that the degrees of freedom are labelled by the continuous variable $\mathbf{x}$;

$$H = \int d^3x \Pi\dot{\phi} - L, \qquad \Pi \equiv \frac{\partial \mathcal{L}}{\partial\dot{\phi}}, \tag{16.29}$$

where $\dot{\phi} \equiv \partial\phi/\partial t$. Using $L = \int \mathcal{L}d^3x$ and Eq. (16.21) this becomes

$$H = \int d^3x \rho(\mathbf{x}), \tag{16.30}$$

where the energy density ρ is given by

$$\rho = \frac{1}{2}\left[\dot{\phi}^2 + |\nabla\phi|^2\right] + V(\phi). \tag{16.31}$$

The vev ϕ_0 of ϕ corresponds to a constant energy density $V(\phi_0)$, which is set to zero because the vacuum energy density corresponds to the cosmological constant. For a homogeneous field,[2]

$$\rho = \frac{1}{2}\dot{\phi}^2 + V(\phi). \tag{16.32}$$

All this is for flat spacetime. According to the equivalence principle, the scalar field equation (16.26) remains valid in curved spacetime, with $\Box$ now given by Eq. (2.27). In the expanding Universe the metric is given by Eqs. (8.1) and (8.2) and Eq. (2.27) gives

$$\ddot{\phi} + 3H\dot{\phi} - a^{-2}\nabla^2\phi + V'(\phi) = 0, \tag{16.33}$$

---

[2] As we will see, this expression remains valid in the expanding Universe.

where $\boldsymbol{\nabla}^2 \equiv \delta^{ij}\partial_i\partial_j$ and $\partial_i \equiv \partial/\partial x_i$. For a homogeneous field,

$$\ddot{\phi} + 3H\dot{\phi} + V' = 0. \qquad (16.34)$$

The middle term slows down the expansion and is sometimes called Hubble drag. For the Fourier components of a free field, we have

$$\ddot{\phi}(\mathbf{k}) + 3H\dot{\phi}(\mathbf{k}) + \left[m^2 + \left(\frac{k}{a}\right)^2\right]\phi(\mathbf{k}) = 0. \qquad (16.35)$$

16.3   ACTION IN CURVED SPACETIME

We will now see how to generalise the action (16.17) to curved spacetime, leading to some powerful results. To do that we need a generalisation of the four-volume element $dV = d^4x$, that is valid for any coordinate choice. According to a standard formula, an immediate generalisation is $dV = Jd^4x$, where J is the determinant of the matrix $\partial x^\mu/\partial x'^\nu$ of the transformation from locally orthonormal coordinates system x'^ν to generic coordinates x^μ.[3] If we regard the right hand side of Eq. (2.4) as a product of three matrices, we arrive at[4] $dV = \sqrt{-g}d^4x$ and

$$S = \int d^4x\sqrt{-g}\mathcal{L}, \qquad (16.36)$$

where instead of Eq. (16.25) we have

$$\mathcal{L} = -\frac{1}{2}g^{\mu\nu}\partial_\mu\phi\partial_\nu\phi - V(\phi). \qquad (16.37)$$

Varying ϕ, one can show that the action principle $\delta S = 0$ gives Eq. (16.26) with $\Box$ given by Eq. (2.27).

The Einstein field equation can also be derived from the action principle. The appropriate action for that purpose is

$$S = \int d^4x\sqrt{-g}\left(\frac{R}{16\pi G} + \mathcal{L}\right), \qquad (16.38)$$

where R is the curvature scalar and $\mathcal{L}$ describes the content of the Universe. (If there is just a scalar field, $\mathcal{L}$ is given by Eq. (16.37).) This is called the Einstein-Hilbert action.

Varying the spacetime curvature, one can show that the action principle gives the Einstein field equation with

$$T_{\mu\nu} = -2\frac{\partial\mathcal{L}}{\partial g^{\mu\nu}} + g_{\mu\nu}\mathcal{L}. \qquad (16.39)$$

---

[3]This is the generalisation, to any number of dimensions, of the two-dimensional formula for the area of a parallelogram.

[4]This expression for dV shows that d^4x is invariant under Lorentz transformations, as I implicitly assumed when discussing Eq. (16.17).

For a scalar field described by Eq. (16.37) it gives

$$T_{\mu\nu} = \partial_\mu\phi\partial_\nu\phi - g_{\mu\nu}\left[\frac{1}{2}g^{\alpha\beta}\partial_\alpha\partial_\beta\phi + V(\phi)\right]. \qquad (16.40)$$

Evaluating $\rho = T_{00}$ with the $\mathcal{L}$ given by Eq. (16.37) and $g_{\mu\nu}$ given by Eqs. (8.1) and (8.2), one verifies that Eq. (16.32) remains valid in the expanding Universe.

EXERCISES

1. Verify Eq. (16.9), giving the Hamiltonian of a harmonic oscillator.

2. Verify Eq. (16.22), the field equation for a scalar field, by inserting the Lagrangian density (16.21) into Eq. (16.20).

3. Read off the metric components corresponding to the Robertson–Walker line element (8.2), and put them into Eq. (2.27) to derive an expression for the d'Alembertian $\square$ in comoving space coordinates. Insert your expression into the general scalar field equation (16.22), to obtain the homogeneous scalar field equation (16.33).

4. Verify that the scalar field energy-momentum tensor (16.40) satisfies $\partial_\mu T^{\mu\nu} = 0$ in flat spacetime. Then verify that it satisfies $D_\mu T^{\mu\nu}$ in curved spacetime, using Eq. (2.17) and the fact that $D_\lambda g_{\mu\nu} = 0$.

Quantum theory of a free field

CONTENTS

This chapter deals with the quantum theory of a free scalar field. We will see how it corresponds to the existence of particles.

17.1   HARMONIC OSCILLATOR

We must first recall the quantum theory of a harmonic oscillator, with the Hamiltonian (16.9). In the quantum theory q and p correspond to operators $\hat{q}$ and $\hat{p}$, which satisfy the commutation relations

$$[\hat{q}, \hat{p}] = i. \tag{17.1}$$

The Hamiltonian operator is

$$\hat{H} = \frac{1}{2}\hat{q}^2 + \frac{1}{2}\omega^2\hat{p}^2. \tag{17.2}$$

In the usual description of the quantum theory, $\hat{q}$ and $\hat{p}$ are constant operators, and the state vector has a time dependence given by the Schrödinger equation. That is called the Schrödinger picture and is usually the most convenient. But for the particular case of the harmonic oscillator, it is easier to go to what is called the Heisenberg picture, in which the state vector is constant while $\hat{q}$ and $\hat{p}$ satisfy the classical equations of motion $\ddot{\hat{q}} = -\omega^2\hat{q}$. In the Heisenberg picture we can write

$$\hat{q}(t) = \frac{1}{\sqrt{2\omega}}\left[e^{-i\omega t}\hat{a} + e^{i\omega t}\hat{a}^\dagger\right]. \tag{17.3}$$

where $\hat{a}$ is a constant operator and the dagger denotes the Hermitian conjugate. The commutator (17.1) is equivalent to

$$[\hat{a}, \hat{a}^\dagger] = 1. \tag{17.4}$$

Using Eq. (17.3) the Hamiltonian is

$$\hat{H} = \left(\hat{a}^\dagger \hat{a} + \frac{1}{2}\right)\omega. \tag{17.5}$$

Eqs. (17.1) and (17.5) give the commutator

$$[\hat{H}, \hat{a}^\dagger] = \hat{a}^\dagger \omega, \tag{17.6}$$

which is equivalent to

$$\hat{H}\hat{a}^\dagger = (1 + \omega)\,\hat{a}^\dagger \hat{H}. \tag{17.7}$$

This means that $\hat{a}^\dagger$, acting on a state with definite energy, increases the energy by an amount ω. Taking the Hermitian conjugates[1] of Eqs. (17.6) and (17.7), we learn that $\hat{a}$ decreases the energy by the same amount. The state $|0\rangle$ satisfying $\hat{a}|0\rangle = |0\rangle$ is the ground state, and from Eq. (17.5) it's energy is $\frac{1}{2}\omega$. The excited states obtained by the action of $\hat{a}^\dagger$ have energies $\frac{3}{2}\omega$, $\frac{5}{2}\omega \cdots$.

Going to a set of oscillators we have

$$[\hat{q}_n, \hat{p}_m] = i\delta_{nm}, \tag{17.8}$$

$$\hat{H} = \frac{1}{2}\sum_n \omega_n^2\left(\hat{q}_n^2 + \hat{p}_n^2\right) \tag{17.9}$$

$$\hat{q}_n(t) = \frac{1}{\sqrt{2\omega}}\left[e^{-i\omega t}\hat{a}_n + e^{i\omega t}\hat{a}_n^\dagger\right], \tag{17.10}$$

and

$$\hat{H} = \sum_n \left(\hat{a}_n^\dagger \hat{a}_n + \frac{1}{2}\right)\omega_n \tag{17.11}$$

$$[\hat{a}_n, \hat{a}_m^\dagger] = \delta_{nm}, \qquad [\hat{a}_n, \hat{a}_m] = 0. \tag{17.12}$$

and

$$\hat{H}\hat{a}_n^\dagger = (1 + \omega_n)\,\hat{a}_n^\dagger \hat{H}. \tag{17.13}$$

The state satisfying $\hat{a}_n|0\rangle$ is the ground state, and $\hat{a}_n^\dagger$ raises the energy by an amount ω_n.

---

[1] Remember that $(AB)^\dagger = B^\dagger A^\dagger$.

17.2 FREE SCALAR FIELD

Corresponding to the field $\phi(\mathbf{x})$ we consider Hermitian operators $\hat{\phi}(\mathbf{x})$, where $\mathbf{x}$ just labels the operator. Adopting the Heisenberg picture, we consider the Fourier series in a box with size L much bigger than any length of interest and write

$$\hat{\phi}(\mathbf{x}, t) = L^{-3} \sum_{\mathbf{k}} \frac{1}{\sqrt{2E_k}} \left[e^{-iE_k t} \hat{a}_{\mathbf{k}} e^{i\mathbf{k}\cdot\mathbf{x}} + e^{iE_k t} \hat{a}_{\mathbf{k}}^{\dagger} e^{-i\mathbf{k}\cdot\mathbf{x}} \right], \qquad (17.14)$$

where the components of $\mathbf{k}$ are positive or negative integers, times $2\pi/L$.[2] I have denoted the angular frequencies by E_k because they will turn out to be the energies of particles. To satisfy the Klein-Gordon equation Eq. (16.22), we need $E_k = \sqrt{m^2 + k^2}$.

The Hamiltonian, given by Eqs. (16.30) and (16.31) with $V = m^2\phi^2/2$, corresponds to the operator

$$\hat{H} = \sum_{\mathbf{k}} \left(L^{-3} \hat{a}_{\mathbf{k}}^{\dagger} \hat{a}_{\mathbf{k}} + \frac{1}{2} \right) E_k \qquad \hat{n}_{\mathbf{k}} \equiv . \qquad (17.15)$$

This is the same as for a set of harmonic oscillators, but $\hat{a}$ in Eq. (17.5) corresponds to $L^{-3/2}\hat{a}_{\mathbf{k}}$ in Eq. (17.11). The commutator is therefore

$$\left[\hat{a}_{\mathbf{k}}, \hat{a}_{\mathbf{k}'}^{\dagger} \right] = L^3 \delta_{\mathbf{k}\mathbf{k}'}. \qquad (17.16)$$

The vacuum state $|0\rangle$ satisfies $\hat{a}(\mathbf{k})|0\rangle = 0$ for all $\mathbf{k}$. Acting n times on the vacuum state with $\hat{a}^{\dagger}(\mathbf{k})$ gives a state with energy $nE(k) + \frac{1}{2}$.

Instead of the Fourier series it is usually more convenient to consider the Fourier integral, corresponding to the limit $L \to \infty$:

$$\left(\frac{2\pi}{L} \right)^3 \sum_{\mathbf{k}} \to \int d^3k \qquad (17.17)$$

$$\hat{a}_{\mathbf{k}} \to \hat{a}(\mathbf{k}) \qquad (17.18)$$

$$\left(\frac{L}{2\pi} \right)^3 \delta_{\mathbf{k}\mathbf{k}'} \to \delta^3(\mathbf{k} - \mathbf{k}'). \qquad (17.19)$$

Then we have

$$\hat{\phi}(\mathbf{x}, t) = \frac{1}{(2\pi)^3} \int d^3k \frac{1}{\sqrt{2E(k)}} \left[e^{-iE(k)t} \hat{a}(\mathbf{k}) e^{i\mathbf{k}\cdot\mathbf{x}} + e^{iE(k)t} \hat{a}^{\dagger}(\mathbf{k}) e^{-i\mathbf{k}\cdot\mathbf{x}} \right], \qquad (17.20)$$

and

$$\hat{H} = \frac{1}{(2\pi)^3} \int d^3k E(k) \left(\hat{a}^{\dagger}(\mathbf{k}) \hat{a}(\mathbf{k}) + \frac{1}{2} L^3 \right), \qquad (17.21)$$

where $E(k) = \sqrt{m^2 + k^2}$. The commutator is

$$[\hat{a}(\mathbf{k}), \hat{a}^{\dagger}(\mathbf{k})] = \delta^3(\mathbf{k} - \mathbf{k}'). \qquad (17.22)$$

---

[2]The sum over $\mathbf{k}$ should be read as a sum over these integers.

17.3   PARTICLES

Reverting to the Fourier series, we have seen that the state obtained by acting on the vacuum n times with $\hat{a}_{\mathbf{k}}$ has energy $[n+\frac{1}{2}]E(k)$, where $E(k) = \sqrt{m^2 + k^2}$. This suggests that $\hat{a}_{\mathbf{k}}^\dagger$ creates a particle with momentum $\mathbf{k}$ and mass m. To justify that interpretation though, I need to show that $\mathbf{k}$ is indeed the momentum of the state.

To do that, I first need the quantum definition of momentum. Taking the momentum to be along say the z axis it has only one component p_z, the momentum operator $\hat{p}_z$ is defined as the operator which generates a shift along the z axis. To be precise, if a system is described by some state vector $|\rangle$, shifting the system along the z axis by an amount Z corresponds to this change in its state vector;

$$|\rangle \to e^{-i\hat{p}_z Z}|\rangle. \tag{17.23}$$

For the quantum theory to work, the same change has to apply to $\hat{A}|\rangle$ where $\hat{A}$ is any operator. We therefore need

$$\hat{A} \to e^{-i\hat{p}_z Z} \hat{A} e^{i\hat{p}_z Z}. \tag{17.24}$$

Taking Z to be infinitesimal, the second equation becomes

$$\hat{A} \to \hat{A} - iZ\left[\hat{p}_z, \hat{A}\right]. \tag{17.25}$$

We need this result for the case that $\hat{A}$ is $\hat{\phi}(\mathbf{x})$. Making that coordinate change in Eq. (17.14) is equivalent to the change

$$\hat{a}_{\mathbf{k}}^\dagger \to (1 - ik_z Z)\,\hat{a}_{\mathbf{k}}^\dagger. \tag{17.26}$$

Comparing Eqs. (17.25) and (17.26), we learn that $[\hat{p}_z, \hat{a}_{\mathbf{k}}^\dagger] = k_z \hat{a}_{\mathbf{k}}^\dagger$, which means that indeed $\hat{a}_{\mathbf{k}}^\dagger$ increases the momentum by an amount $\mathbf{k}$. It follows that acting n times on the vacuum with $\hat{a}_{\mathbf{k}}^\dagger$ creates a state with momentum $n\mathbf{k}$. As we saw earlier, $\hat{a}_{\mathbf{k}}^\dagger$ also increases the energy by an amount $\sqrt{m^2 + k^2}$. This justifies the statement that $\hat{a}_{\mathbf{k}}^\dagger$ creates a particle with momentum $\mathbf{k}$, energy $E(k)$ and mass m. For this reason, $\hat{a}_{\mathbf{k}}^\dagger$ is called a creation operator.

17.4   VACUUM STATE

In the expression (17.21) for $\hat{H}$, L is the size of the box that was used for the Fourier series. It doesn't disappear when we take the limit of large L to go to the Fourier integral, but that makes sense because $\hat{H}$ is the operator corresponding to the energy. The energy is proportional to L and the energy divided by L^3 is the energy density.

According to Eq. (17.21), the vacuum state has energy density

$$\rho_{\text{vac}} = \frac{2\pi}{(2\pi)^3} \int_0^\infty k^2 E(k)\,dk. \tag{17.27}$$

This is infinite, but a quantum field theory is valid only for k below some 'ultra-violet cutoff'. Including only those k, and denoting the cutoff by Λ_{UV}, the vacuum energy density due to a scalar field becomes[3]

$$\rho_{\text{vac}} = \frac{\pi/2}{(2\pi)^3}\Lambda_{UV}^4. \tag{17.28}$$

This, along with the contributions of other fields, makes up the quantum part of the cosmological constant that was mentioned in Chapter 5. The fact that ρ_{vac} is of order Λ_{UV}^4 could have been anticipated on dimensional grounds, since we are working in natural units and Λ_{UV} is the only relevant quantity.

Using Eqs. (17.20) and (17.22) with

$$\int e^{i(\mathbf{k}-\mathbf{k}')\cdot\mathbf{x}}d^3x = (2\pi)^3\delta^3(\mathbf{k}-\mathbf{k}'), \tag{17.29}$$

one finds the mean square deviation of ϕ from its vev;

$$\langle[\widehat{\delta\phi}(\mathbf{x})]^2)\rangle = \pi\Lambda_{UV}^2. \tag{17.30}$$

It is independent of $\mathbf{x}$ because the vacuum is invariant under translations. Again, the order of magnitude is to be expected from the dimension.

17.5 PLANCK SCALE

Having discussed the vacuum fluctuation of a scalar field, I am in a position to describe what is called the Planck scale. The Planck scale is the regime where the present framework of physics, provided by General Relativity and particle physics, ceases to work.

To arrive at the Planck scale, we need to consider the energy, length and time units that can be formed from the fundamental constants. They are called the Planck energy, the Planck length and the Planck time.[4]

$$E_P = \sqrt{\hbar c^5/8\pi G} = 2.43 \times 10^{18}\,\text{GeV} \tag{17.31}$$
$$L_P = \sqrt{\hbar G/c^3} = 1.62 \times 10^{-35}\,\text{m} \tag{17.32}$$
$$T_P = L_P/c = 5.39 \times 10^{-44}\,\text{s}. \tag{17.33}$$

To understand the significance of these quantities, we can begin by considering the average energy E within a sphere of radius L, that is generated by the quantum fluctuation of a scalar field with negligible mass. Working in natural units, averaging over the sphere is roughly removing Fourier components with $k > 2\pi/L$. Using Eqs. (17.19) and (17.22), this gives $E(L) \sim L^{-1}$ (as expected, since L is the only

---

[3]The mass has been set equal to zero, which is an excellent approximation because the mass has to be much less than Λ_{UV} for the field theory to be valid.

[4]The factor 8π in the E_P is inserted according to a common convention.

relevant quantity). If $E(L)$ is big enough, the quantum fluctuation will form a black hole.

To see how big that is, one needs to know that with the units restored, a black hole will form if matter with mass M is confined within a sphere whose radius R is roughly equal to GM/c^2.[5] Going to natural units, we conclude that a black hole will form if $L \sim GE(L)$. Since $E(L)$ is roughly $1/L$ in natural units, that corresponds to $L \sim \sqrt{G}$, and restoring the units that becomes $L \sim L_P$.

According to quantum field theory then, it makes no sense to talk about the energy within a region that is smaller than the Planck length. This places two important restrictions on the validity of quantum field theory.

The first restriction concerns the collision of two particles. According to quantum field theory, the typical distance probed by a head-on collision between relativistic particles with energy E is $L \sim 1/E$ in natural units. If E is bigger than the Planck energy, L is less than the Planck distance and quantum theory ceases to make sense. That is a pity because some cosmic ray particles do have much bigger energy and they do occasionally collide. Nobody knows what happens when such a collision occurs. (The energies of the collisions at LHC are only about 10^6 MeV by the way and it would be completely impossible to build a collider with the Planck energy.)

The second restriction concerns the phenomenon of black hole evaporation. Assuming that the size of the a black hole is much bigger than the Planck length, one can use quantum field theory to demonstrate that the black hole would evaporate by emitting particles.[6] The evaporation rate is far too small to be of interest for known black holes. But tiny black holes might be formed in the early Universe, whose evaporation might be observable. As the evaporation proceeds, the black hole becomes smaller and eventually it becomes as small as the Planck length. Then the quantum field theory calculation no longer applies, and nobody knows what happens next. Some people think that it will continue to evaporate till there's nothing left and others think that it will stop evaporating, but nobody knows.

EXERCISES

1. Verify Eqs. (17.4), (17.5), and (17.6).

2. Use Eq. (17.7) to calculate the effect of $\hat{a}^\dagger$ on a state with definite energy.

3. Verify Eq. (17.11).

4. Verify Eq. (17.26).

5. Verify Eq. (17.30).

---

[5]To be precise, this happens if the surface area of the sphere is $4\pi R$, with $R = 2GM/c^2$. One cannot be precise about the radius, because Euclidean geometry doesn't apply and the radius tends to infinity as the black hole limit is reached.

[6]This was done in 1974 by Steven Hawking, building on previous work.

Inflation

CONTENTS

Now we deal with the era of inflation, which is believed to set the initial conditions for the Big Bang. Inflation is defined as an era during which $aH = \dot{a}$ increases with time, corresponding to repulsive gravity. This is in contrast to the situation during the Big Bang when aH decreases with time corresponding to attractive gravity.

The definition of inflation can be stated in an equivalent way using

$$(aH)^{\cdot} = \dot{a}H + a\dot{H} = a\left(H^2 + \dot{H}\right). \tag{18.1}$$

From this, we see that inflation may be defined as an era during which

$$\dot{H} < -H^2. \tag{18.2}$$

To describe inflation one uses natural units. In those units G has the dimension E^{-2} and it is convenient to work, not with G itself but with E_{P} defined by Eq. (17.31). In natural units, energy and mass are identical and it is usual to denote E_{P} by M_{P} calling it the Planck mass:

$$M_{\mathrm{P}} \equiv \frac{1}{\sqrt{8\pi G}} = 2.436 \times 10^{18}\,\mathrm{GeV} \qquad \text{natural units.} \tag{18.3}$$

18.1 TWO STAGES OF INFLATION

Inflation is able to set the initial conditions for the Big Bang, more or less independently of the conditions at the beginning of inflation. For that to happen though, inflation must begin early enough. To be precise, it must begin when the region that will become the observable Universe is well inside the Hubble distance.

This requirement for the beginning of inflation can be formulated in terms of the value of aH. To do that we should remember first that the distance across the observable Universe is at present roughly the Hubble distance $1/H_0$. As we go back in time, the size of the region that will become the observable Universe is therefore roughly a/H_0. The requirement that this region be well inside the Hubble distance is therefore $a/H_0 \ll 1/H$. Remembering that $a = 1$ at present, that in turn is equivalent to requiring that aH is much less than its present value.

We have then two stages of inflation. The first stage ends when aH achieves its present value. As we will see, the first stage is supposed to create an unperturbed Universe that is in accordance with observation. The second era is supposed to create, on top of that, cosmological perturbations whose statistical properties are in accordance with observation. Only the second stage can be probed directly by observation, and I will call it observable inflation.

The first stage will inevitably do its job, provided that it lasts long enough. Let us see how that is. The first stage can make the observable Universe practically homogeneous, because the repulsive gravity will drive energy *from* regions of higher energy density *into* regions with lower energy density (the opposite of what gravity does during the Big Bang). As for the isotropy, it can be shown that expansion always reduces the anisotropy. Enough expansion can therefore make the expansion practically isotropic even if it was not so initially. Finally, the first stage can make the $\dot{a}/K \gg 1$ (as observation requires) even if that is not so initially, because $\dot{a}$ increases during inflation.

For the first stage to make the observable Universe nearly homogeneous, it is important that the observable Universe during that era is inside the Hubble distance. This is because it takes about a Hubble time for anything to travel a Hubble distance. During the first stage there is plenty of time for things to travel across the region that will become the observable Universe so that it can be made homogeneous. Afterwards, taking the expansion of the Universe into account, that would not be possible.

18.2   AMOUNT OF OBSERVABLE INFLATION

According to Eq. (18.2), H has to be slowly varying during observable inflation. Taking it to have some constant value H_{inf}, we have during inflation $a \propto \exp(H_{\text{inf}}t)$. The number N of Hubble times of observable inflation is then equal to $\ln(a_1/a_2)$ where 1 denotes the beginning of observable inflation and 2 denotes its end. Using this result, I now calculate N under various assumptions about the subsequent evolution of the scale factor.

The simplest thing is to assume radiation domination, which lasts until the onset of the present matter-dominated era. For simplicity I ignore the cosmological constant which makes only a small difference to the results. Also, I assume a sudden transition from radiation- to matter domination at the epoch $z_{\text{eq}} = 3390$ when the two are equal. Since observable inflation begins when aH has its present value,

$$1 = \frac{a_1 H_{\text{inf}}}{a_0 H_0} = e^{-N} \frac{a_2 H_{\text{inf}}}{a_{\text{eq}} H_{\text{eq}}} \frac{a_{\text{eq}} H_{\text{eq}}}{a_0 H_0}. \tag{18.4}$$

During radiation domination we have $\rho \propto a^{-4}$ and $a \propto t^{1/2}$. During matter domination we have $\rho \propto a^{-3}$ and $a \propto t^{2/3}$. During both eras we have $H \propto 1/t$. Using these results, Eq. (18.4) gives

$$N = 62 - \ln \frac{10^{16}\,\text{GeV}}{\rho_{\text{inf}}^{1/4}}. \tag{18.5}$$

Here, $\rho_{\text{inf}} = 3M_{\text{P}}^2 H_{\text{inf}}^2$ is the energy density during inflation. As we shall see, observation requires $\rho_{\text{inf}}^{1/4} \lesssim 10^{16}\,\text{GeV}$, which implies $N < 62$. On the other hand, inflation must end before the beginning of the known history. As we saw in Chapter 7 that requires $\rho_{\text{inf}}^{1/4} \gtrsim 1\,\text{MeV}$, implying $N \gtrsim 18$.

As we will see, a more realistic assumption is that inflation is followed by a matter-dominated era, which gives way to radiation domination only around the "reheating epoch" at which thermal equilibrium is established.[1] One then has

$$N = 62 - \ln \frac{10^{16}\,\text{GeV}}{\rho_{\text{inf}}^{1/4}} - \frac{1}{3} \ln \frac{\rho_{\text{inf}}^{1/4}}{\rho_{\text{reh}}}, \tag{18.6}$$

where reh denotes the reheating epoch. This reduces N for a given value of ρ_{inf}. It is usually supposed though that N is in the upper part of its possible range $18 \lesssim N \lesssim 62$.

18.3 SLOW-ROLL INFLATION

The energy density during inflation cannot come from particles, because these generate attractive gravity. It must come instead from one or more scalar fields. In the simplest case it comes entirely from a single scalar field, called the inflaton.

After K becomes negligible, the Friedmann equation with Eq. (16.32) gives

$$3M_{\text{P}}^2 H^2 = V(\phi) + \frac{1}{2}\dot{\phi}^2. \tag{18.7}$$

Differentiating it using Eq. (16.34) gives the useful expression

$$2M_{\text{P}}^2 \dot{H} = -\dot{\phi}^2. \tag{18.8}$$

These equations are exact. To get a viable scenario for inflation, we need what is called the slow-roll approximation. The starting point for this approximation is the statement that H varies slowly, corresponding to

$$|\dot{H}| \ll H^2. \tag{18.9}$$

With Eqs. (18.7) and (18.8) this is equivalent to

$$3M_{\text{P}}^2 H^2 \simeq V.. \tag{18.10}$$

---

[1] The term reheating is used because inflation was originally assumed to be preceded by an era of thermal equilibrium. There is no motivation for such an era, which is nowadays regarded as unlikely.

If the latter approximation is valid, it is reasonable to hope that its low derivatives with respect to t will also be valid. Using Eq. (18.8), the first derivative of Eq. (18.10) reads

$$3H\dot{\phi} \simeq -V'(\phi).$$ (18.11)

This is equivalent to neglecting the first term of the exact field equation (16.33):

$$|\ddot{\phi}| \ll 3H|\dot{\phi}|.$$ (18.12)

Eq. (18.12) states that $\dot{\phi}$ doesn't change much in one Hubble time. It follows from Eqs. (18.10) and (18.11) that

$$\epsilon(\phi) \ll 1 \qquad \text{where} \quad \epsilon \equiv \frac{M_{\text{Pl}}^2}{2}\left(\frac{V'}{V}\right)^2.$$ (18.13)

Going further, we can hope that the derivative of the approximation (18.11) is also valid:

$$\ddot{\phi} \simeq -\frac{\dot{H}}{H}\dot{\phi} - \frac{V''\dot{\phi}}{3H}.$$ (18.14)

Comparing with the exact equation (16.34), we see that Eqs. (18.10), (18.11), and (18.14) imply

$$|\eta(\phi)| \ll 1 \qquad \text{where} \quad \eta \equiv M_{\text{P}}^2 \frac{V''}{V} \simeq \frac{V''}{3H^2}.$$ (18.15)

The approximations (18.10), (18.11) and (18.14) constitute the slow-roll approximation. The slow-roll approximation implies the conditions (18.13) and (18.15) on the potential, which we will call **flatness conditions**.

Within the slow-roll approximation, the number of Hubble times before the end of inflation is given by

$$N(\phi) \simeq \frac{1}{M_{\text{P}}^2} \int_\phi^{\phi_{\text{end}}} \frac{V}{V'} d\phi.$$ (18.16)

Slow-roll inflation typically ends when one of the flatness conditions is violated. Then, ϕ moves towards a minimum of the potential around which it oscillates. The potential near the minimum will typically be of the form (16.28). In that case, inflation gives way to a matter dominated era, in which the universe is populated by inflaton particles with negligible random motion. At some stage the inflaton particles decay into particles moving with speed close to c. We then have a radiation-dominated era, which will typically persist until the onset of the matter dominated era that is now being ended by the onset of the cosmological constant.

18.4 DARK ENERGY

To end this chapter, I leave inflation and return to the contribution to the present energy density that is called the cosmological constant. Although present observations are compatible with the hypothesis that this contribution is time independent, that may not be so in the future. With that in mind, the contribution is often called dark energy.

If it is time dependent, the dark energy cannot belong to the vacuum, because the vacuum is something that doesn't change with time. It might instead come from the potential of some scalar field. Adopting that hypothesis, the dark energy is called quintessence, and the field responsible for it is called the quintessence field. The quintessence potential should satisfy conditions similar to those required for the inflaton potential, so that the dark energy density varies slowly.

The idea of quintessence, is that the dark energy density would be zero if the quintessence field had its vacuum value. That sounds reasonable, but the problem is that there is no known reason why the energy density of the vacuum should be zero. In particular, there is no known symmetry of quantum field that would ensure that. Nevertheless, the quintessence idea is taken very seriously by the cosmology community and there is intense interest in future observations that may show a change in the dark energy density as we go back in time.

EXERCISES

1. Use Eqs. (6.6) and (8.7) to show that inflation requires $p/\rho < 3$ if that ratio is constant.

2. Verify Eq. (18.5).

3. Verify Eq. (18.13).

4. Taking H to be constant during inflation, and assuming radiation domination thereafter, show that the number of Hubble times between horizon exit for a scale and the end of inflation is equal to the number of Hubble times between the end of inflation and horizon entry.

5. Verify Eqs. (18.13), (18.15), and (18.16).

Perturbations from inflation

CONTENTS

In this chapter we will see how the era of observable inflation can generate the primordial curvature perturbation ζ, that exists at the beginning of the known history and is supposed to be the origin of all structure in the universe. We will also see how it generates a gravitational wave background, that may be observable in the future. I will use natural units and the slow-roll results of Chapter 18.

19.1 GENERATING THE INFLATON PERTURBATION

The curvature perturbation is generated by the perturbation $\delta\phi$ of the inflaton field. The inflaton perturbation in turn is generated by the vacuum fluctuation. In this section we see how that happens.

Using the linear approximation $\delta V' = V'' \delta\phi$, Eq. (16.33) gives for a Fourier component of $\delta\phi$

$$\delta\ddot{\phi}(\mathbf{k}) + 3H\delta\dot{\phi}(\mathbf{k}) + \left(\frac{k}{a}\right)^2 \delta\phi(\mathbf{k}) + V''(\phi)\delta\phi(\mathbf{k}) = 0. \tag{19.1}$$

During inflation, $aH = \dot{a}$ increases with time, and for each cosmological wavenumber there is an epoch when $a/k = 1/H$. After inflation, aH decreases with time, and as shown in Table 11.1 there is an epoch when again $a/k = 1/H$. The latter epoch is called the epoch of horizon entry, and I will call the former epoch the epoch of **horizon exit**.[1]

To see how $\delta\phi$ is generated from the vacuum fluctuation, we need only consider

---

[1]During the Big Bang, the term horizon entry is used because the Hubble distance is roughly the biggest distance that can have been travelled since the beginning of the Big Bang (the horizon). During inflation, one still calls the Hubble distance the distance to the horizon, though there is actually no limit to the distance that light can travel during inflation.

the era around horizon exit, corresponding to $k/a \sim H$. Then, by virtue of the slow-roll condition $\epsilon \ll 1$ we can take H to be constant. For definiteness I will set H equal to its value at the epoch $k/a = H$, which I will denote by H_k. By virtue of the slow-roll condition $|\eta| \ll 1$ we can also drop the last term of Eq. (19.1). After these things have been done, it becomes convenient to replace $\delta\phi$ by $\varphi = a\delta\phi$, and to replace t by conformal time η, which satisfies $d\eta = dt/a$. Using $a \propto \exp(H_k t)$ we can write $\eta = -1/aH_k$. Then Eq. (19.1) becomes

$$\frac{d^2\varphi(\mathbf{k})}{d\eta^2} + \left[k^2 - \frac{2}{\eta^2} \right] \varphi(\mathbf{k}) = 0. \tag{19.2}$$

For the operator corresponding to φ we can write, analogously with Eq. (17.20),

$$\hat{\varphi}(\mathbf{x}, \eta) = \frac{1}{(2\pi)^3} \int d^3k \left[u(k, \eta)\hat{a}(\mathbf{k})e^{i\mathbf{k}\cdot\mathbf{x}} + u^*(k, \eta)\hat{a}^\dagger(\mathbf{k})e^{-i\mathbf{k}\cdot\mathbf{x}} \right] \tag{19.3}$$

$$= \frac{1}{(2\pi)^3} \int d^3k \hat{\varphi}(\mathbf{k}, \eta)e^{i\mathbf{k}\cdot\mathbf{x}} \tag{19.4}$$

$$\hat{\varphi}(\mathbf{k}, \eta) \equiv u(k, \eta)\hat{a}(\mathbf{k}) + u^*(k, \eta)\hat{a}^\dagger(-\mathbf{k}), \tag{19.5}$$

with u a solution of Eq. (19.2) that I choose as

$$u(k, \eta) = \frac{e^{-ik\eta}}{\sqrt{2k}} \frac{(k\eta - i)}{k\eta}. \tag{19.6}$$

The early-time regime $k\eta \gg 1$ corresponds to $k/a \gg H$. In this regime we can work in a locally inertial frame, with size L over a time interval T such that

$$a/k \ll T \ll 1/H \qquad a/k \ll L \ll 1/H. \tag{19.7}$$

(The upper limits ensure that the expansion of the Universe can be ignored, which means that a can be taken to be constant. Then the comoving coordinates x^i can be taken to be the physical coordinates of the locally inertial frame, because the actual physical coordinates $a(t)x^i$ differ only by a factor which can be taken to be constant and therefore just represents a change in the distance unit.) The equivalence principle then requires that we reproduce the flat spacetime quantum theory of Chapter 12. The solution (19.6) reproduces Eq. (17.20) for $\delta\phi = \varphi/a$, with $m = 0$ which means that in the early time regime we are dealing with a massless scalar field. As we are dealing with the era of observable inflation, there are no particles which means that the state vector of the system is the vacuum state $|0\rangle$, satisfying

$$\hat{a}(\mathbf{k})|0\rangle = 0|0\rangle. \tag{19.8}$$

for all $\mathbf{k}$. This state is time independent because we are working in the Heisenberg picture.

We are going to see that the field perturbation $\varphi(\mathbf{k}, t)$ can generate the primordial curvature perturbation $\zeta(\mathbf{k})$. For that to be possible though, $\varphi(\mathbf{k}, t)$ should be a classical as opposed to a quantum object. In other words, there should be a state vector which (to a good approximation) is at all times an eigenvector of the operator $\hat{\varphi}(\mathbf{k}, t)$, the eigenvalue being $\varphi(\mathbf{k}, \eta)$. Looking at Eq. (19.3) that does not seem possible, because $\hat{\varphi}(\mathbf{k}, t)$ is a different operator at different times, and a state vector cannot in general be the eigenvector of two or more different operators. But a few Hubble times after horizon exit, we have $k\eta \ll 1$ which gives to a good approximation

$$u(k, \eta) = -\frac{i}{\sqrt{2k}}\frac{1}{k\eta}.$$ (19.9)

This is purely imaginary and gives

$$\hat{\varphi}(\mathbf{k}, \eta) = u(k, \eta)\left[\hat{a}(\mathbf{k}) - \hat{a}^\dagger(-\mathbf{k})\right],$$ (19.10)

which means that the operator $\hat{\varphi}(\mathbf{k}, \eta)$ now changes only by a numerical factor. As a result, there *does* exist a state vector which is an eigenvector of $\hat{\varphi}(\mathbf{k}, \eta)$ at all times.[2] This means that $\varphi(\mathbf{k}, \eta)$ becomes a classical quantity a few Hubble times after horizon exit, and so does $\delta\phi = \varphi/a$. Including only the classical Fourier components, we arrive at a classical quantity $\delta\phi(\mathbf{x})$, whose Fourier components have a/k much bigger than the horizon.

The state vector defined by Eq. (19.8) is a superposition of the state vectors that correspond to definite values of the observables $\varphi(\mathbf{k}, \eta)$. We identify that superposition with the ensemble that is invoked to define the statistical properties of the cosmological perturbations, and we suppose that our universe is a typical member of the ensemble. The ensemble average is to be identified with the expectation value in the state defined by Eq. (19.8), and using Eqs. (17.22), (19.6), (19.3) and (19.9) we find the spectrum of $\delta\phi = \varphi(\mathbf{k})/a$ a few Hubble times after horizon exit:

$$\mathcal{P}_{\delta\phi}(k) = \left(\frac{H_k}{2\pi}\right)^2.$$ (19.11)

Although I have focussed on the inflaton field perturbation, the calculation applies to any field whose potential satisfies $|V''(\phi)| \ll H^2$. Such a field is called a light field.

19.2 GENERATING ζ

We have seen that the inflaton perturbation acquires a classical perturbation during inflation, with the spectrum (19.11). We have also noticed that more generally, any light field will acquire the same perturbation. The primordial curvature perturbation might be generated by the inflaton perturbation or by the perturbation of some other light field. In this section I describe the first possibility, and then make some remarks about the second possibility.

---

[2] I am focussing on a single $\mathbf{k}$ to keep the language simple, but I actually mean that there exists a state vector which is an eigenvector of all of the operators $\hat{\varphi}(\mathbf{k}, \eta)$ that have $k\eta \gg 1$. That is possible because the commutator (17.22) implies that the $\hat{\varphi}(\mathbf{k}, \eta)$ with different $\mathbf{k}$ commute.

In discussing the inflaton field perturbation $\delta\phi$ I have taken spacetime to be unperturbed. One can show that this is a good approximation if we define the inflaton perturbation $\delta\phi$ on the slicing of uniform a. The curvature perturbation $\zeta = \delta a/a$ is instead defined on the slicing of uniform energy density. Through Eq. (11.5), this gives $\zeta = (d\ln a/dt)\delta t$ and $\delta\phi = -\dot\phi\delta t$, where δt goes from the first slicing to the second. Eliminating δt we find

$$\zeta = -H\delta\phi/\dot\phi. \tag{19.12}$$

As all Fourier components of $\delta\phi$ have wavelength much bigger than the horizon, we can use the separate universe assumption to discuss the evolution of ζ. As seen in Section 11.4, ζ is constant provided that the pressure is determined by the energy density. That will be the case if the inflaton field is the only relevant field, because according to the separate universe assumptions its value at each location then determines the entire future evolution of the universe at that location.[3]

Using the slow-roll result for $\dot\phi$ and the spectrum (19.11), the spectrum $\mathcal{P}_\zeta$ of ζ is

$$\mathcal{P}_\zeta(k) = \left(\frac{H}{2\pi}\right)^2 \frac{9H^4}{V'^2}. \tag{19.13}$$

Using again the slow-roll results, we can then calculate a quantity called the spectral index, which is denoted by n_s;

$$n_s(k) - 1 \equiv \frac{1}{\mathcal{P}_\zeta(k)}\frac{d\mathcal{P}_\zeta(k)}{dk} = 2\eta - 6\epsilon. \tag{19.14}$$

The first expression is the definition of $n_s(k)$. In the second expression η and ϵ are evaluated at the epoch when $aH = k$, and the expression is obtained using the slow-roll results. For a typical potential, $n_s(k)$ doesn't vary much over cosmological scales, and taking it to be a constant it becomes the n_s defined by Eq. (11.29).

This scenario for generating ζ gives non-Gaussianity that is probably too small ever to observe. If instead ζ is generated from the perturbation of a light field different from the inflaton, ζ can have significant non-gaussianity. One typically has

$$\zeta = \zeta_g + \frac{3}{5}f_{NL}\zeta_g^2, \tag{19.15}$$

where ζ_g is Gaussian and f_{NL} is some constant. (The factor 3/5 is a convention of historical origin.) The most commonly considered scenario for generating ζ in this way, called the curvaton scenario, gives in its simplest form $f_{NL} = 5/4$. Data from the Planck satellite give at the time of writing $f_{NL} = 0.8 \pm 5$, but the accuracy of future observations may be sufficient to confirm or rule out the simplest curvaton prediction.

---

[3]To be precise it determines the evolution for as long as a/k remains well outside the horizon. But ζ is anyway defined only in that regime.

19.3 GENERATING GRAVITATIONAL WAVES

Inflation also generates gravitational waves from the vacuum fluctuation. Perturbing the action (16.38) to second order in $g_{\mu\nu}$, with $\mathcal{L}$ set to zero, one finds that the action for each of the amplitudes $h_{+,\times}$ defined in Eq. (3.43) is the same as for a massless scalar field perturbation $\delta\phi_{+,\times}$ given by

$$\delta\phi_{+,\times} = \sqrt{2}M_{\mathrm{P}}h_{+,\times}. \tag{19.16}$$

The gravitational wave spectrum $\mathcal{P}_h$ is defined by

$$4\langle h_+(\mathbf{k})h_+(\mathbf{k}')\rangle = 4\langle h_\times(\mathbf{k})h_\times(\mathbf{k}')\rangle = (2\pi)^3\frac{2\pi^2}{k^3}\mathcal{P}_h(k)\delta^3(\mathbf{k}-\mathbf{k}'). \tag{19.17}$$

Using Eqs. (11.12), (19.11), and (19.17), we conclude that a few Hubble times after horizon exit

$$\mathcal{P}_h(k) = \frac{8}{M_{\mathrm{P}}^2}\left(\frac{H_k}{2\pi}\right)^2. \tag{19.18}$$

According to Eq. (19.18) it is slowly varying.

The time dependence of $h_{+,\times}$ is that of a massless scalar field:

$$\ddot{h}_{+,\times}(\mathbf{k}) + 3H\dot{h}_{+,\times}(\mathbf{k}) + (k/a)^2 h_{+,\times} = 0. \tag{19.19}$$

It follows that $h_{+,\times}$ becomes constant soon after horizon exit during inflation, and remains so until the approach of horizon entry. The spectrum (19.18) also remains constant. The relative magnitude of the gravitational waves is defined by the parameter

$$r \equiv \frac{\mathcal{P}_h(k_0)}{\mathcal{P}_\zeta(k_0)}. \tag{19.20}$$

The absence of a detection so far gives the upper bound $r < 0.11$.

19.4 CONSTRAINING THE INFLATON POTENTIAL

Observation constrains the form of the inflaton potential, if we require that the curvature perturbation is generated from the inflaton perturbation (optional) and that the gravitational wave background is below the observational limit (compulsory). From a theoretical viewpoint, the best-motivated potential $V(\phi)$ leading to slow-roll inflation is the sinusoidal potential:

$$V(\phi) = 1 + \cos(\phi/M), \tag{19.21}$$

with $M \gg M_{\mathrm{P}}$. The motivation is not very strong though, and the sinusoidal potential leads to somewhat complicated expressions. To illustrate the use of the slow-roll expressions, I will therefore use the simple (though unmotivated) potentials

$$V \propto \phi^p, \tag{19.22}$$

with p any positive number, not necessarily integer. The slow-roll parameters are

$$\epsilon = \frac{p^2}{2}\frac{M_{\mathrm{P}}^2}{\phi^2}, \qquad \eta = p(p-1)\frac{M_{\mathrm{P}}^2}{\phi^2}. \tag{19.23}$$

Inflation ends at $\phi_{\mathrm{end}} \simeq pM_{\mathrm{P}}$, and the potential can be modified at smaller ϕ without affecting the predictions (assuming of course that the modification does not lead to further inflation).

The number of Hubble times before the end of inflation is related to ϕ by

$$\phi \simeq \sqrt{2N(\phi)p}\,M_{\mathrm{P}}. \tag{19.24}$$

Using Eqs. (19.23) and (19.24), one finds $r = 4p/N$ and $1 - n_{\mathrm{s}} = (2+p)/2N$, where N is the number of Hubble times of observable inflation. Using the observed value $1 - n_{\mathrm{s}} = 0.04$, the second relation gives $p = 0.08N - 2$. Putting this into the first relation, and invoking the observational bound $r < 0.11$, one finds $p < 1.1$ and $N < 40$. The value of N is rather low but not impossible, which means that the power-law potential is not yet ruled out by observation.

EXERCISES

1. Verify Eqs. (19.13) and (19.14).

2. Verify Eqs. (19.23) and (19.24).

Prehistory of the Big Bang

CONTENTS

It is natural to ask what might have happened between the end of inflation and the beginning of the known history, which one might call the prehistory. In this chapter I give the simplest scenario and comment on alternatives.

When inflation ends, the inflaton field begins to oscillate corresponding to the presence of inflaton particles. This marks the beginning of the Big Bang. As the inflaton field is nearly homogeneous, the particles have negligible random motion. In the terminology of Chapter 5, we are dealing with a matter dominated era.[1] The matter-dominated era ends when the matter particles decay, to produce a radiation-dominated Universe.[2] The radiation particles presumably undergo collision processes, which quickly establish thermal equilibrium at a temperature that is determined by the energy density. This marks the beginning of what is called the Hot Big Bang.

20.1 PREHISTORY WITH THE STANDARD MODEL

I will first describe the prehistory of the Hot Big Bang, assuming that the Standard Model accounts for everything, even though we know that this cannot be completely correct. Let us suppose first that it begins when kT is bigger than $100\,\text{GeV}$ or so. In this case, collision processes occur with enough kinetic energy to create all of the elementary particles described by the Standard Model.[3] There are no hadrons,

---

[1] If inflation involves more than one field, more than one field might be oscillating corresponding to the presence of more than one particle species. They would still constitute matter.

[2] To be precise, that is when matter domination finally ends. Before the particles decay, the oscillating field can lose energy by a process called preheating, which involves the field itself and has nothing to do with particles. The particles created by preheating can constitute radiation, but the density of such radiation will fall more quickly than the energy density of the oscillating field and the latter is expected to dominate by the time that the inflaton particles decay.

[3] The quarks and leptons actually have zero mass when kT is much bigger than $100\,\text{GeV}$, because high temperature changes the Higgs potential, so that its minimum value corresponds to a zero

nuclei or atoms because if any of these objects managed to form they would quickly be blown apart by collisions.

As we saw in Chapter 7, the state of thermal equilibrium is determined by the temperature, and the densities of the three conserved quantities which are zero or very small. As a result, the elementary particles will have with high accuracy the generalized blackbody distributions defined by Eqs. (7.11) and (7.10).

As the Universe expands, it cools and the distance between the particles increases. To understand what happens next, one needs to know that a quark or gluon can exist, only if there are other quarks and gluons within a distance of about 10^{-15} m. This is called confinement, and it ceases to be true when kT falls to about 100 MeV. The quarks and gluons are then replaced by hadrons, which are made out of quarks. Of the hadrons, the proton is stable, and the neutron has a longish half-life (15 minutes). But all other hadrons have very short lifetimes; they decay, and are not re-created because their rest energies are bigger than the kinetic energies of the photons and leptons that might have done the job. At about the same time, the Higgs bosons decay for the same reason, as do Z and W bosons and the μ and τ leptons listed in Table A.5. We are left with just the particle species that are present at the beginning of the known history of the Universe.

If kT is less than 100 GeV at the start of the Hot Big Bang, the history is the same except that it begins at the appropriate value of kT. For instance, if kT is less than 100 MeV at the start, there are never any isolated quarks and gluons, only hadrons.

20.2 CREATING THE CDM

The prehistory that I have given assumes that before quark confinement, there are more u and d quarks than anti-u and anti-d quarks. If the numbers were equal quark confinement would produce equal numbers of protons and anti-protons (and equal numbers of neutrons and anti-neutrons) which would annihilate. As a result, the Universe would contain no ordinary matter. It may be that the excess of quarks over anti-quarks is produced by the decay of the inflaton, in which case the excess would not require any modification of the prehistory. More usually, it is supposed that the excess is created later through some modification of the prehistory. To keep things simple though, I won't discuss the latter possibility, since it is after all optional.

What I do need to discuss, is the creation of CDM. That's compulsory because the Standard Model contains no species that could be the CDM. In contrast with the situation for the inflaton, there *are* extensions of the Standard Model that include a CDM candidate, and yet are motivated by considerations that have nothing to do with CDM. There are two such extensions and I discuss them in turn.

According to one extension, the CDM is a particle species called the axion,

Higgs field. But when kT falls to about 100 GeV the Higgs potential reverts to its usual form and the quarks and leptons acquire their mass.

which corresponds to the oscillation of a scalar field called the axion field. This extension is motivated by the fact that, according to the Standard Model, the neutron should have a dipole moment. The magnitude of the dipole moment depends on a parameter of the Standard Model, and the fact that no dipole has yet been observed requires that the parameter be less than 10^{-12}. The axion extension of the Standard Model can explain why the parameter is so small, but I will not deal with that. Instead I will explain how the axion might be important for cosmology and astronomy.

There are two possible scenarios for producing axion CDM. In one scenario, the axion field exists already during inflation, when it is homogeneous except for a small perturbation generated from the vacuum fluctuation. After inflation, the axion field oscillates around its vacuum value corresponding to the presence of axions. In this scenario the adiabatic condition is not exactly satisfied, because the axion field has a perturbation on cosmological scales that is different from the inflaton field's perturbation, but the departure from the adiabatic condition can be too small to observe.

In the other scenario, the axion field comes into existence only after inflation. In this scenario the adiabatic condition is almost exactly satisfied on cosmological scales. On very small scales though, the CDM density perturbation is much bigger than the one corresponding to the adiabatic condition. This scenario predicts the existence of very small pure CDM halos. They have not been observed, but that may be because they have merged like the bigger pure CDM halos mentioned in Chapter 15. Both scenarios can therefore be in accordance with observation.

The interaction of axions with Standard Model particles is very weak, in accordance with the fact that the axions can be the CDM, but it exists and it may allow the detection of axion CDM. Also, the interaction means that axions, if they exist, are emitted by the sun whether or not they are the CDM. Experiments are underway to detect both CDM axions and solar axions.

So much for the axion extension of the Standard Model. The other extension, invoking what is called supersymmetry, is motivated by what is called the hierarchy problem. The hierarchy problem is the fact that the Standard Model gives a quantum contribution to the mass of the Higgs boson that is far bigger than the observed quantity. That is similar to the situation for the cosmological constant and it has the same implication; there must be another "classical" contribution to the Higgs mass, which almost exactly cancels the quantum contribution. The supersymmetry scenario seeks to avoid the hierarchy problem, by giving each Standard Model species a 'superpartner'. The quantum contribution to the Higgs boson mass then becomes about the same as the mass of a typical superpartner. The hierarchy problem would be completely removed if the superpartners were only about as heavy as the Higgs boson. That is ruled out by the fact that superpartners have not been produced at the LHC. There remains the possibility though, that the superpartners exist and are only say ten or a hundred times heavier than the Higgs boson. Then the difference between the classical and quantum contributions would be 10% or 1% which might not be regarded as a problem.

With supersymmetry, the prehistory of the Hot Big Bang is the same until after the electroweak transition, except for the presence of the superpartners. After the electroweak transition though, the superpartners decay except for the lightest superpartner which is stable and is the CDM candidate.

The generic name for *any* CDM particle with properties similar to the lightest superpartner is Weakly Interacting Massive Particle, abbreviated as WIMP. As with the axion, experiments are underway to detect WIMP CDM through its interaction with Standard Model particles. The interaction might allow the direct detection of WIMPs, or their indirect detection through electromagnetic radiation emitted when WIMPs in our galaxy collide.[4] It might also allow WIMPs to be produced at the LHC. If the WIMPs are light, they may have significant random motion corresponding to what is called Warm Dark Matter. That might be detected by comparing the structure of galaxies with the outcome of computer simulations of their formation.

20.3 GUT COSMIC STRINGS

In this book I am focussing on ideas about the early Universe that are known to be correct, or at least are not ruled out by observation. Of course there have been plenty of other ideas, that seemed promising at first but were later ruled out by observation. Here is the story of a famous example.

It starts in the late 1970's, when an extension of the Standard Model called Grand Unified Theory (GUT) was proposed. Around 1980, it was realised that such an extension would allow the creation of what are called cosmic strings. A cosmic string is a concentration of energy within a tube, which is so narrow so that it can for most purposes be regarded as a line. The energy within the string comes from fields that are far from their vacuum values.

The GUT cosmic strings would form a network, consisting of both long lines and small loops. It was proposed that galaxies formed, not in regions where the cosmic gas was denser than average, but instead around cosmic string loops. That idea seemed reasonable because a loop attracts matter towards it by the force of gravity, just like an over-dense region of the gas. Unfortunately, it failed to account for observation (in particular for the CMB anisotropy) and was abandoned in the early 1990's.

There remained the possibility that the network of GUT cosmic strings could exist. If so, a pair of strings would occasionally intersect and it was shown that this would lead to the emission of a gravitational waves. It was also shown that it would cause a variation in the observed period of pulsars, whose magnitude was calculated. The failure to detect a variation of the required magnitude in the 2000s ruled out the GUT cosmic strings.

It is still possible that a network of cosmic strings exist, similar to the GUT

---

[4]Radiation is observed coming from the centre of our galaxy, for which WIMP collisions provide one of the most plausible explanations — but unfortunately not the only one.

strings but with a lower mass per unit length. In particular, there could be a network consisting of the strings of string theory. Such a network might be detected in the future, but it would have nothing to with galaxy formation.

The second of the two axion scenarios mentioned earlier is also associated with cosmic strings. These, called axionic cosmic strings, are quite different from GUT strings and will have disappeared before the beginning of the known history of the Universe.

20.4   WHAT MIGHT FUTURE OBSERVATIONS FIND?

I've now finished my account of cosmology, as it stands today. It remains to ask what the future may hold. The most dramatic cosmological discovery would be to detect the gravitational wave background. That would be a very strong indication that inflation occurred, because the gravitational wave background includes wavelengths not much less than the size of the observable Universe that could hardly have been generated by any other mechanism. Moreover, it would tell us that inflation occurred at a very early stage, leaving plenty of time for things to happen between the end of inflation and the beginning of the known history.

A less dramatic discovery would be a departure from the adiabatic condition, which relates the primordial energy density perturbations to the primordial curvature perturbation. As we saw in Section 20.2, such a departure is possible if, for example, the CDM consists of axions. Unfortunately though, there is no particular reason to think that the departure will be at a level that will be observable in the near future. A discovery that the dark energy varies with time would also be very interesting. It would mean that it doesn't come from the vacuum but instead, presumably, from some 'quintessence' field.

There is the possibility of finding out more about the CDM as mentioned earlier. It could be that the CDM comes from two or more particle species. Conceivably, it could be that the CDM doesn't consist of particles at all, but of sizeable objects. These couldn't be made out of ordinary matter. They might be black holes formed before the beginning of the known history though the mass of such black holes must lie within a very narrow range.

Even if none of that happens, we will for sure learn more about the spectrum of the primordial curvature perturbation. We will probably find a smooth departure from the simple form that currently fits observation, at least if ζ is generated from the inflaton perturbation. But we might instead find a departure that is not smooth — a bump or a dip in the curve that represents the spectrum. That would indicate that something special happened during inflation, at the epoch during inflation when the corresponding wavelength became as big as the Hubble distance.

EXERCISES

1. Use Eq. (7.10) to verify that the spacing between quarks and gluons is 10^{-5} m when kT is about 100 MeV.

2. For GUT cosmic strings, the energy per unit length in natural units is about 10^{14} GeV. Using these units, calculate the energy of a GUT cosmic string whose length is equal to the Hubble distance during an epoch of radiation domination, in terms of the age of the Universe. Verify that this energy is much less than the energy of the radiation within a sphere whose diameter is the Hubble distance.

Appendix

CONTENTS

A.1   SPECIAL FUNCTIONS

This section deals with the spherical harmonics $Y_{\ell m}$ that are used to expand the CMB anisotropy $\Delta T/T$, and the weighted spherical harmonics $_{\pm}Y_{\ell m}$ that are used to expand the CMB polarisation specified by $Q_{\pm}$.

The spherical harmonics are defined by

$$Y_{\ell m} = \left[\frac{2\ell+1}{4\pi} \frac{(l-m)!}{(l+m)!} \right] P_{\ell}^{m}(\cos\theta)e^{im\phi}, \qquad (A.1)$$

where the associated Legendre function is defined by[1]

$$P_{\ell}^{m}(x) = (-1)^{\ell} \frac{(1-x^2)^m}{2!\ell!} \frac{d^{\ell+m}}{dx^{\ell+m}} (1-x^2)^{\ell}. \qquad (A.2)$$

They are a complete orthonormal set, satisfying

$$\int Y_{\ell m}(\mathbf{e}) Y_{\ell'm'}^{*}(\mathbf{e}) \, d\Omega = \delta_{\ell\ell'} \delta_{mm'}, \qquad (A.3)$$

where $d\Omega \equiv d\cos\theta \, d\phi$. They satisfy $Y_{\ell m}^{*} = (-1)^m Y_{\ell,-m}$. Under a reversal of the direction (θ, ϕ), corresponding to a parity transformation,

$$Y_{\ell m} \to Y_{\ell m}(\pi-\theta, -\phi) = (-1)^{\ell} Y_{\ell,-m}(\theta, \phi). \qquad (A.4)$$

A rotation of the polar coordinate system leads to a unitary transformation for the $Y_{\ell m}$, with no mixing between different ℓ;

$$Y_{\ell m}(\theta, \phi) \to \sum_{m'} U_{mm'}(\ell) Y_{\ell m'}(\theta', \phi'). \qquad (A.5)$$

---

[1]This is a widely used convention. Some authors differ by a factor $(-1)^{\ell}$, and/or by a factor $(-1)^m$ for negative m.

Coming now to the $Q_\pm$, we have to remember that the polarization tensor P_{ij} is defined using Cartesian coordinates $\{x^2, x^2\}$ where 1 is the direction of increasing θ in some spherical coordinate system $\{r, \theta, \phi\}$ and 2 is the direction of increasing ϕ. If we go to a different spherical coordinate system, we rotate these directions through some angle ψ, taken to be clockwise about the r direction. This transforms P_{ij} and hence $Q_\pm$.

As P_{ij} is a tensor it behave under rotations like $A_i B_j$ where A_i and B_j are vectors. The vectors in turn behave like the coordinates x^i, which means that under a rotation through an angle ψ about the 3 direction we have

$$
\begin{aligned}
A_1' &= A_1 \cos\psi + A_2 \sin\psi \\
A_2' &= -A_1 \sin\psi + A_2 \cos\psi
\end{aligned}
\tag{A.6}
$$

and similarly for B_i. Using trigonometric identities it follows that

$$
\begin{aligned}
Q' &= Q\cos(2\psi) + U\sin(2\psi) \\
U' &= -Q\sin(2\psi) + U\cos(2\psi).
\end{aligned}
\tag{A.7}
$$

This gives

$$
Q_\pm' = \exp(\pm 2i\psi)Q_\pm.
\tag{A.8}
$$

The functions $_\pm Y_{\ell m}$, in terms of which $Q_\pm$ are expanded, are defined so that they acquire the appropriate factor $\exp(\pm 2i\psi)$ when the spherical polar coordinates are changed. It can be shown that this requirement gives (up to a factor which is chosen to satisfy Eq. (A.10))

$$
_\pm Y_{\ell m} = 2\sqrt{\frac{(\ell-2)!}{(\ell+2)!}} \left(\frac{\partial}{\partial\theta} \pm \frac{im}{\sin\theta}\right)\left(\frac{\partial}{\partial\theta} \pm \frac{im}{\sin\theta}\right) Y_{\ell m}.
\tag{A.9}
$$

The functions are orthonormal,

$$
\int {}_\pm Y_{\ell m}(\mathbf{e}) {}_\pm Y^*_{\ell'm'}(\mathbf{e})\, d\Omega = \delta_{\ell\ell'}\delta_{mm'}
\tag{A.10}
$$

$$
\int {}_\pm Y_{\ell m}(\mathbf{e}) {}_\mp Y^*_{\ell'm'}(\mathbf{e})\, d\Omega = 0.
\tag{A.11}
$$

A.2 COSMOLOGICAL PERTURBATIONS

I here provide the derivations of Eq. (12.4) and Eqs. (12.10)–(12.13). To arrive at Eq. (12.4) we need to generalise Eq. (11.5) in two ways. First, we must allow a generic change of coordinates, corresponding to spatial shifts δx^i as well as the time shift δx^0. Second, we must replace f by the metric tensor $g_{\mu\nu}$, whose transformation under a coordinate change is given by Eq. (3.21). The coordinate change in our case is

$$
x'^\alpha = x^\alpha + \delta x^\alpha(x_0, x_1, x_2, x_3),
\tag{A.12}
$$

giving

$$\frac{\partial x'^{\alpha}}{\partial x^{\mu}} = \delta_{\mu}^{\alpha} + \partial_{\mu}\delta x^{\alpha} \tag{A.13}$$

whose first-order inverse is[2]

$$\frac{\partial x^{\mu}}{\partial x'^{\alpha}} = \delta_{\alpha}^{\mu} - \partial_{\alpha}'\delta x^{\mu}. \tag{A.14}$$

Using Eqs. (2.11) and (A.14) we have to first order at a given spacetime position

$$\delta g'_{\mu\nu} - \delta g_{\mu\nu} = -g_{\alpha\nu}\partial_{\mu}\delta x^{\alpha} - g_{\mu\alpha}\partial_{\nu}\delta x^{\alpha}. \tag{A.15}$$

Allowing also for the change in spacetime position we have (generalising Eq. (11.5))

$$\delta g'_{\mu\nu} - \delta g_{\mu\nu} = -g_{\alpha\nu}\partial_{\mu}\delta x^{\alpha} - g_{\mu\alpha}\partial_{\nu}\delta x^{\alpha} - \delta x^{\alpha}\partial_{\alpha}g_{\mu\nu}. \tag{A.16}$$

We want to apply this result to Eq. (12.1). As we want to consider only the scalar metric perturbations, we write $\delta x^{i} = -i(k^{i}/k)\delta x$ for the spatial components. Then Eq. (A.15) gives

$$A' = A - (\delta\eta) - aH\delta\eta \tag{A.17}$$

$$B' = B + (\delta x) + k\delta\eta \tag{A.18}$$

$$D' = D - \frac{k}{3}\delta x - aH\delta\eta \tag{A.19}$$

$$E' = E + k\delta x. \tag{A.20}$$

If Eq. (12.1) is not already of the form (12.4), we can set E to zero by the choice of δx and then set B to zero by the choice of $\delta\eta$.

To derive Eqs. (12.10)–(12.13), I will until the end of the calculation use t as the time coordinate instead of η. The first step is to calculate the perturbation in Γ_i^{jk} using its definition (2.14). Working with Fourier components one finds[3]

$$\delta\Gamma_{00}^{0} = \dot{\Psi} \tag{A.21}$$

$$\delta\Gamma_{0i}^{0} = \delta\Gamma_{i0}^{0} = ik_{i}\Psi, \tag{A.22}$$

$$\delta\Gamma_{ij}^{0} = \delta\Gamma_{ji}^{0} = -\delta_{ij}a^{2}\left[2H\left(\Phi + \Psi\right) + \dot{\Phi}\right] \tag{A.23}$$

$$\delta\Gamma_{j0}^{i} = \delta\Gamma_{0j}^{i} = -\delta_{ij}\dot{\Phi} \tag{A.24}$$

$$\delta\Gamma_{jk}^{i} = -i\Phi\left(\delta_{ij}k_{k} + \delta_{ik}k_{j} - \delta_{jk}k_{i}\right). \tag{A.25}$$

We also need the perturbations in the metric tensor, defined by Eqs. (12.5)–(12.9). Inserting these and the $\delta\Gamma_{\mu\nu}^{\alpha}$ into Eq. (3.8) and working to first order, one arrives at the perturbed continuity equations (12.10) and (12.11).

---

[2]To check this, use both expressions to work out $(\partial x'^{\alpha}/\partial x^{\mu})(\partial x^{\mu}/\partial x'^{\beta})$ to first order, finding that it is indeed δ_{β}^{α}.

[3]Since δ_{ij} and k_{i} live in flat space, raising their indices has no effect and I choose always the lower position.

The next step is to calculate the perturbations in R_{00}, R_{ij} and R using Eqs. (2.42)–(2.44). One finds

$$\delta R_{00} = -\frac{k^2}{a^2}\Psi + 3\ddot{\Phi} + 3H\left(\dot{\Psi} + 2\dot{\Phi}\right) \tag{A.26}$$

$$\delta R_{ij} = \delta_{ij}\left[-2\left(2a^2H^2 + a\ddot{a}\right)(\Psi + \Phi) - a^2H\left(6\dot{\Phi} + \dot{\Psi}\right) - a^2\ddot{\Phi} - k^2\Phi\right]$$
$$+ k_ik_j(\Psi - \Phi), \tag{A.27}$$

$$\delta R = -12\left(H^2 + \frac{\ddot{a}}{a} - \frac{k^2}{6a^2}\right)\Psi - 6\ddot{\Phi} - 6H\left(\dot{\Psi} + 4\dot{\Phi}\right) - 4\frac{k^2}{a^2}\Phi. \tag{A.28}$$

Putting these and the metric perturbations into the perturbed Einstein field equation one finds to first order

$$k^2\Phi + 3H\left(\dot{\Phi} + H\Psi\right) = -4\pi Ga^2\delta\rho \tag{A.29}$$

$$k^2\left(\dot{\Phi} + H\Psi\right) = 4\pi Ga^2(\rho + P)V \tag{A.30}$$

$$\ddot{\Phi} + H\left(\dot{\Psi} + 2\dot{\Phi}\right) + \left(2\frac{\ddot{a}}{a} - H^2\right)\Psi + \frac{k^2}{3}(\Phi - \Phi) = 4\pi Ga^2\delta P \tag{A.31}$$

$$k^2(\Phi - \Psi) = 8\pi Ga^2 P\Pi \tag{A.32}$$

As we are already using the two continuity equations, which follow from the Einstein equations, the latter need only provide us with two more equations. In the text these have been chosen as Eq. (12.13) and Eq. (12.12). After using the Friedman equation to relate H^2 with ρ, one sees that Eq. (12.13) is Eq. (A.32), and that Eq. (12.12) follows from Eqs. (A.29) and (A.30).

A.3 CONSTANTS, PARAMETERS AND SYMBOLS

1 m	=	$5.068 \times 10^{15}\,\mathrm{GeV}^{-1}\hbar c$
1 s	=	$1.519 \times 10^{24}\,\mathrm{GeV}^{-1}\hbar$
1 kg	=	$5.608 \times 10^{26}\,\mathrm{GeV}/c^2$
1 Joule	=	$6.242 \times 10^{-5}\,\mathrm{GeV}$
1 K	=	$8.618 \times 10^{-14}\,\mathrm{GeV}/k_{\mathrm{B}}$

Table A.1 **Conversion from natural units to MKS and kelvin**

Reduced Planck constant	$\hbar$	=	$1.055 \times 10^{-34} \, \text{J s}$
Speed of light	c	=	$2.998 \times 10^{8} \, \text{m/s}$
Newton's constant	G	=	$6.672 \times 10^{11} \, \text{m}^3 \, \text{kg}^{-1} \, \text{s}^{-2}$
Reduced Planck mass	M_{Pl}	=	$2.436 \times 10^{18} \, \text{GeV}/c^2$
Boltzmann constant	k_{B}	=	$1.381 \times 10^{-23} \, \text{J/K}$
Thomson cross section	σ_{T}	=	$6.652 \times 10^{-29} \, \text{m}^2$
Solar mass	$M_\odot$	=	$1.99 \times 10^{30} \, \text{kg}$
Megaparsec	$1 \, \text{Mpc}$	=	$3.086 \times 10^{22} \, \text{m}$
Year	$1 \, \text{yr}$	=	$3.156 \times 10^{7} \, \text{s}$

Table A.2 **Some numbers in MKS units**

Hubble time	$1/H_0$	=	$14.5 \pm 0.1 \, \text{Gyr}$
CMB temperature	T_0	=	$2.725 \pm 0.001 \, °\text{K}$
CDM energy density fraction	Ω_{c}	=	0.264 ± 0.001
Baryon energy density fraction	Ω_{b}	=	0.049 ± 0.001
Cosmological constant fraction	Ω_Λ	=	0.68 ± 0.01
Spectrum of ζ	$10^9 \mathcal{P}_\zeta(k_0)$	=	2.21 ± 0.07
Spectral tilt	$n_{\text{s}} - 1$	=	-0.035 ± 0.005

Table A.3 **Fundamental cosmological parameters**. To calculate the known history of the homogeneous Universe one needs (in addition to the fundamental constants and the relevant Standard Model parameters) five cosmological parameters. These can be chosen to be the ones in the first five rows, defined at the present epoch. To describe the inhomogeneity one needs A and n_{s}, which as shown in Eq. (11.29) specify the spectrum $\mathcal{P}_\zeta(k)$ of ζ. The values of the parameters are chosen so that the calculated CMB spectrum C_ℓ agrees with measurements made in the Planck spacecraft, and are taken from the preprint "Planck 2015 results. XIII Cosmological parameters" by the Planck Collaboration. With that choice, there is adequate agreement with all other observations.

Hubble distance	cH_0^{-1}	$= 4.46\,\text{Gpc}$
Present temperature	$k_B T_0$	$= 2.36 \times 10^{-4}\,\text{eV}$
Energy density	ρ_0	$= 4.76\,\text{GeV m}^{-3}$
CMB energy density	Ω_γ	$= 5.4 \times 10^{-5}$
Neutrino energy density	Ω_ν	$= 3.7 \times 10^{-5}$
Horizon distance	x_{hor}	$= 1.40 \times 10^4\,\text{Mpc}$
Baryon-to-photon ratio	n_B/n_γ	$\equiv \eta = 6.1 \times 10^{-10}$
Redshift at last scattering	$z_{\ell s}$	$= 1100$
Redshift at equality	z_{eq}	$= 3390$
Hubble length at equality	$c(a_{\text{eq}} H_{\text{eq}})^{-1}$	$= 97.9\,\text{Mpc}$

Table A.4 Derived cosmological parameters. These are calculated from the fundamental cosmological parameters and the known history. The first five entries refer to the present epoch. The neutrino energy density fraction is calculated under the pretence that the neutrinos are massless. The last two entries refer to the epoch when the radiation and matter energy densities are equal. The last entry is the *comoving* Hubble length.

particle	γ (photon)	gluon	Z	$W_\pm$	Higgs boson
rest energy	—	—	$91,190$	$80,420$	$125,000$

particle	u	d	c	s	t	b
rest energy	2.5	5.0	1270	95	$173,000$	4240

particle	e (electron)	μ (muon)	τ	ν_1	ν_2	ν_3
rest energy	0.511	106	1780	5×10^{-8}	9×10^{-9}	?

neutron	proton
939.6	938.3

Table A.5 Rest energies The tables show (i) the bosons, (ii) the quarks, (iii) the leptons and (iv) the proton and neutron. The rest energies are in MeV. The mass of a particle in kilograms is equal to 1.79×10^{-30} times its rest energy in MeV. The neutrino rest energies are known to be very small. According to the most likely interpretation of the observations, two of them have the values indicated and the third has a much smaller value.

	charge	baryon number	L_e	L_μ	L_τ
u, c & t quarks	2/3	1/3	0	0	0
d, s & b quarks	−1/3	1/3	0	0	0
proton	1	1	0	0	0
neutron	0	1	0	0	0
electron	−1	0	1	0	0
μ	−1	0	0	1	0
τ	−1	0	0	0	1
ν_e	0	0	1	0	0
ν_μ	0	0	0	1	0
ν_τ	0	0	0	0	1
photon	0	0	0	0	0
Higgs boson	0	0	0	0	0
gluon	0	0	0	0	0
Z	0	0	0	0	0
W_+	1	0	0	0	0

Table A.6 **Conserved quantities** This table shows the amount of each conserved quantity carried by the elementary particles and the nucleons. Anti-particles carry equal and opposite amounts. The electric charges are in units of the proton charge and the other charges are dimensionless.

Symbol	Page	Definition
$a(t)$	42	Scale factor of the Universe
$a_{\ell m}$	108	Multipole of $\Delta T/T$
C_ℓ	110	Spectrum $\langle |a_{\ell m}|^2 \rangle$ of CMB anisotropy
f	55	Occupation number
$g_{\mu\nu}$ (g_{ij})	71	Spacetime (space) metric tensor
h_{ij}	28	Gravitational wave amplitude
H (H_0)	43	Hubble parameter $\dot{a}/a$ (present value)
k $(\mathbf{k})$	79	Comoving wavenumber (wave vector)
L $(\mathcal{L})$	125	Lagrangian (Lagrangian density)
n	57	Number density
n_{s}	92	Spectral index of ζ
N	140	Hubble times of observable inflation
P	21	Pressure
$\mathbf{p}$ (p) (p^μ)	4	Momentum (magnitude of) (4-momentum)
$\mathcal{P}_g$	88	Spectrum of a perturbation g
r	149	Tensor fraction $\mathcal{P}_h/\mathcal{P}_\zeta$
$T^{\mu\nu}$	21	Energy momentum tensor
$\mathbf{v}$	21	Fluid velocity
V	80	Fluid velocity scalar
$V(\phi)$	128	Scalar field potential
x (x^μ) (x^i)	65	Comoving distance (spacetime coordinates) (space coordinates)
w	97	Ratio P/ρ for a fluid
z	49	Redshift
δ	79	Density contrast $\delta\rho/\rho$
ϵ	142	Slow-roll flatness parameter $\frac{1}{2}M_{\mathrm{P}}^2(V'/V)^2$
ζ	91	Primordial curvature perturbation
η	142	Slow-roll flatness parameter $M_{\mathrm{P}}^2 V''/V$
η	96	Conformal time $d\eta = dt/a$
$\eta_{\mu\nu}$	3	Metric tensor (Minkowski coordinates)
ρ (ρ_0)	21	Energy density (of present Universe)
Π	96	Anisotropic stress scalar
ϕ	127	Scalar field
Φ	79	Newtonian peculiar gravitational potential
Φ, Ψ	96	Metric perturbations
φ	146	Conformal inflaton field perturbation $a\delta\phi$
Ω_{s}	54	Present ρ_{s}/ρ of species 's"

Table A.7 This table shows the meaning of some symbols with their place of definition.

Index

Printed in the United States
by Baker & Taylor Publisher Services